Satellite Radar Interferometry

Subsidence Monitoring Techniques

Remote Sensing and Digital Image Processing

VOLUME 14

For other volumes published in this series go to www.springer.com/series/6477

Satellite Radar Interferometry

Subsidence Monitoring Techniques

By
V.B.H. (Gini) Ketelaar

Delft University of Technology,
The Netherlands

Springer

V.B.H. (Gini) Ketelaar
Nederlandse Aardolie Maatschappij,
Assen, The Netherlands

Cover figure: Fig. 2.6 from this book.

Responsible Series Editor: Freek D. van der Meer

Remote Sensing and Digital Image Processing   ISSN 1567-3200

ISBN 978-1-4020-9427-9   e-ISBN 978-1-4020-9428-6

Library of Congress Control Number: 2008944289

Printed on acid-free paper.

9 8 7 6 5 4 3 2 1

springer.com

Preface

Since the start of hydrocarbon production in the Netherlands, measurement campaigns have been performed to measure the resulting subsidence, to which gas and oil companies in the Netherlands are legally obliged. The majority of the gas fields in the Netherlands, including the Groningen gas field, are operated by Nederlandse Aardolie Maatschappij B.V. (NAM). Different subsidence measurement techniques (leveling, GPS) have been utilized since the 1960s. Synchronously, geodetic estimation methodologies have been developed to estimate subsidence due to hydrocarbon production from the measurements, in which the Delft Institute of Earth Observation and Space Systems (DEOS) has been closely involved. Since the 1990s, satellite radar interferometry (InSAR) as a deformation monitoring technique has developed. However, the situation in the Groningen area is challenging (temporal decorrelation, rural areas, atmospheric disturbances, small deformation rates—several mm/year—over a large spatial extent). In 2003, the project 'Fundamenteel Onderzoek Radar Interferometrie' was approved (Regeling Technologische Samenwerking), which enabled a four year PhD research to investigate the feasibility of InSAR for monitoring subsidence due to hydrocarbon production, in cooperation between Delft University of Technology and NAM. This book describes the results of this scientific research, that is directly coupled to the practical demand for subsidence monitoring techniques. It covers the topic in a generic way: both precision and reliability of InSAR as a measurement technique and the estimation of earth surface deformation in the presence of multiple deformation causes are addressed.

Acknowledgements

This book is the result of a PhD research, that has been performed in cooperation between Delft University of Technology and NAM, between the radar remote sensing group of Prof. Hanssen (DEOS, DUT) and the onshore surveys team of Lammert Zeijlmaker (NAM). The project has been supported by SenterNovem, agency of the Dutch Ministry of Economic Affairs. All ERS and Envisat SAR data were kindly provided by the European Space Agency (ESA) for Category 1 project 2724, "InSAR deformation analysis of subsidence in the Groningen gas field, the Netherlands". Many visualizations in this book have been created using Matlab®, which has been utilized for numerous computations.

I am very grateful to my promotors Prof. Hanssen and Prof. Teunissen, and Lammert Zeijlmaker of NAM for giving me the opportunity to perform this scientific research with such direct practical implementations. I have experienced a very pleasant and open working environment both at Delft University of Technology and at NAM. In particular, I would like to thank Prof. Hanssen for the support during the entire research period, the detailed review of this book and many valuable suggestions. I also greatly appreciate the feedback and critical comments I have received from Prof. Teunissen and the members of the examination committee: Prof. Klees,

Prof. Kroonenberg, Prof. Rocca, Dr. Duquesnoy, and Dr. Smit. Series editor Freek van der Meer and Petra Steenbergen from Springer are acknowledged for publishing this work in the *Remote Sensing and Digital Image Processing* series.

Furthermore, I would like to thank all (former) members of the radar remote sensing group at Delft University of Technology for the pleasant working environment: Joaquin Munoz Sabater, Freek van Leijen, Petar Marinkovic, Yue Huanyin, Swati Gehlot, Rossen Grebenitcharsky, Zbigniew Perski, Ayman Elawar, Liu Guang, Miguel Caro Cuenca, Mahmut Arikan, Jia Youliang, Frank Kleijer, Gert Jan van Zwieten and Shizhuo Liu, and Bianca Cassee during her MSc graduation project. Special thanks go to Freek van Leijen and Petar Marinkovic, whose work has definitely speeded up the obtained results for subsidence monitoring in the Groningen region, for the open attitude during my entire PhD research period, and the inspiring discussions. I would also like to thank Bert Kampes for his quick response to many questions. Thanks go as well to Alireza Amiri-Simkooei for assistance with Variance Component Estimation, Roderik Lindenbergh for assistance with geostatistics, Ria Scholtes for the administrative support, and to all other members of the Mathematical Geodesy and Positioning (MGP) department of Delft University of Technology for the nice working environment. I would like to thank Hans Garlich and Joop Gravesteijn for the assistance with leveling the corner reflectors. Furthermore, I would like to thank Adriaan Houtenbos for the useful hints he has given me. Going back to the initial contact at the start of my PhD research, I would like to thank Frank Kenselaar for reacting enthusiastically when contacting him after spending four and a half years working in the industry. At NAM, I have received a lot of useful feedback from the subsidence monitoring team of Lammert Zeijlmaker and the geomechanics team of Dirk Doornhof. I have appreciated the cooperation with Simon Schoustra, Wilfred Veldwisch and Stefan Kampshoff, and would like to thank Onno van der Wal for all subsidence prognoses. Finally, I would like to thank my parents Gert and Marijke, brother Joris, and Arnoud, for their patience and support.

Audience

The research described in this book investigates the applicability of satellite radar interferometry (InSAR) for deformation monitoring, in particular subsidence due to hydrocarbon extraction. It covers the subject in a generic way, from the precision and reliability of InSAR as a measurement technique to the estimation of the deformation signal of interest in the potential presence of multiple deformation causes. It provides an overview of the Persistent Scatterer InSAR (PSI) theory, and subsequently focuses on the accuracy of the parameter estimates. For the reliability assessment of InSAR deformation estimates, which is essential for operational use, the multi-track datum connection procedure is introduced. The presented methodologies are demonstrated in an integrated way for the entire northern part of the Netherlands and a part of Germany (covering $\sim$15.000 km^2) using time series of ERS and Envisat acquisitions. The capabilities of PSI for wide-scale monitoring of subsidence rates of several millimeters per year in rural areas are shown. Furthermore, it is demonstrated that the temporal observation density of PSI improves the insight in hydrocarbon reservoir behavior. The reader is assumed to have a background in geosciences and to be familiar with basic radar interferometry concepts. The book is designed for both researchers and the industry, since it translates the research results into the consequences for the operational use of InSAR for subsidence monitoring.

Readers who are interested in a geological background of the Groningen gas reservoir and the prediction of subsidence at ground level are referred to Chap. 2. For the theoretical background of PSI and its precision and reliability, the reader is recommended to focus on the Chaps. 3, 4, and 5. If one has a background in PSI and is looking for the specific application for subsidence monitoring due to gas extraction in the Netherlands, the reader is referred to Chap. 6, preceded by Chap. 5, which addresses the reliability assessment methodology for PSI deformation estimates. Readers who are most interested in the operational use of PSI for monitoring subsidence due to hydrocarbon production are referred to Chap. 7. To conclude, Chap. 8 addresses the potential of PSI for improving knowledge on reservoir behavior.

Contents

Summary

The start of hydrocarbon production in the 1960s in the northeastern part of the Netherlands has resulted in subsidence of the ground level, which has been estimated from periodic leveling campaigns. Although leveling is a precise and reliable technique for subsidence monitoring, it is labor intensive, expensive and poses a safety risk since measurements are taken along roads. Hence, the application of satellite radar interferometry (InSAR) is investigated for subsidence monitoring, coupled with the potential improvement of reservoir behavior monitoring due to the InSAR observation frequency. The main focus lies on the Groningen gas field, which has a diameter of $\sim$30 kilometers, at $\sim$3 kilometers below surface. Complicating factors for the application of InSAR for subsidence monitoring in the Groningen area are surface changes in time due to its agricultural character (temporal decorrelation), atmospheric disturbances, and the low subsidence rates ($<$1 cm/year) over a large spatial extent. Hence, the applicability of Persistent Scatterer InSAR (PSI) is investigated. PSI utilizes objects with a coherent phase behavior in time for the estimation of deformation and other phase contributions. Since the subsidence monitoring period exceeds the lifetime of a satellite (5–10 years), multiple sensors are required: ERS-1, ERS-2 and Envisat phase observations have been used.

Both the presence of PS in rural areas and the accuracy of PSI deformation estimates have been investigated. The PS density varies from 0–10 PS per km^2 in rural areas to more than 100 PS per km^2 in urban areas. Approximately 80% of the Groningen subsidence area is covered with at least one PS per km^2. The quality assessment of PSI for monitoring subsidence due to hydrocarbon production has two components: the precision and reliability of the measurement technique, and the relation of the deformation estimates to the deformation signal of interest (the idealization precision). The stochastic model of PSI has been validated in a controlled corner reflector experiment, using independent leveling measurements. The estimated precision for ERS-2 and Envisat double-difference displacements is 3.0 and 1.6 mm (1-sigma) respectively. The correlation coefficient between the Envisat and leveling double-difference displacements is 0.94. The precision of displacements for natural PS (objects in the terrain) in the Groningen area varies from $\leq$3 mm in urbanized areas to 3–7 mm (1-sigma) in rural areas, where the distances between neighboring PS are larger.

Only the fractional phase of a PS is observed; the number of integer phase cycles is unknown. Since the integer ambiguities are included as unknowns in the system of equations, there is no redundancy in the estimation procedure. Hence, a testing procedure on outliers and model errors can not be performed in the parameter estimation of a single arc spanned by two PS. However, under the assumption of ambiguity resolution success rates of 1, the effect of imperfections in the mathematical model of PSI has been assessed. Inaccuracies in the azimuth sub-pixel position can lead to additional errors of $\sim$0.5 mm/year in the PS displacement rate (velocity) estimates. Random orbit errors with a standard deviation of 5 and 8 cm in radial and across-track direction respectively can lead to velocity errors up to ~1 mm/year between near and far range. Regarding the stochastic model, the possibilities of variance component estimation (VCE) have been investigated. Moreover, a reference independent quality measure has been proposed, the Dilution of Precision.

A multi-track datum connection procedure has been developed to perform a reliability assessment, if success rates of 1 cannot be guaranteed. Multi-track datum connection utilizes overlapping independent tracks that redundantly observe the same deformation signal. The Groningen subsidence bowl is (partly) observed by six ERS tracks (adjacent and cross-heading). After datum connection, the standard deviation of the PS velocity estimates is less than 1 mm/year for 70% of the clusters of nearby multi-track PS. Moreover, multi-track deformation estimates have been utilized to decompose deformation along line-of-sight into vertical and horizontal movements.

Subsidence due to hydrocarbon production can be contaminated by other deformation regimes, such as foundation instabilities and shallow compaction. The idealization precision for the estimation of the deformation signal of interest may be improved by utilizing the physical PS properties and by using a-priori knowledge on the spatio-temporal behavior of the deformation signal of interest. PS characterization to improve the idealization precision is based on the assumption that direct reflections from (well-founded) buildings are the most suitable targets for the estimation of deformation due to deep subsurface displacements. PS heights, Envisat Alternating Polarization observations, and the PS reflection pattern as a function of viewing geometry have been utilized to select PS that represent direct reflections from elevated targets. Case studies in two areas show a shift towards velocity estimates of a lower magnitude after PS selection, but this shift is not significant (<0.5 mm/year). Since subsidence due to hydrocarbon production is the common deformation regime, this implies that it can be estimated from PS that are selected based on spatial correlation, provided that the majority of the buildings in an area are well-founded. It is recommended to evaluate this condition in all subsiding areas.

The deformation estimates from PSI and leveling campaigns have been cross-validated, taking the accuracy of both measurement techniques into account. The correlation coefficient between the displacement rates of both techniques is 0.94, comparable to the correlation coefficient of displacements in a controlled corner reflector experiment (0.94), and comparable to the correlation coefficient of displacement estimates from repeated leveling campaigns (~0.94–0.97). Moreover, the spatio-temporal density can be applied for monitoring reservoir behavior, e.g., the uplift due to underground gas storage is captured by PSI. It can be concluded that PSI has reached the maturity to be operationally used for monitoring subsidence due to gas extraction in the northern part of the Netherlands—stand-alone or, in specific cases, in concert with significantly reduced leveling campaigns or GPS.

Nomenclature

List of Acronyms and Abbreviations

ALD	Azimuth Look Direction
APS	atmospheric phase screen
DEM	Digital Elevation Model
DIA	Detection Identification Adaptation
DOP	Dilution of Precision
ERS	European Remote Sensing Satellite
ESA	European Space Agency
Envisat	Environmental Satellite
FFT	Fast Fourier Transform
GIS	Geographical Information System
GPS	Global Positioning System
ILS	Integer Least-Squares
InSAR	Interferometric Synthetic Aperture Radar
LAMBDA	Least-squares AMBiguity Decorrelation Adjustment
LOS	line of sight
NAM	Nederlandse Aardolie Maatschappij B.V.
NAP	Normaal Amsterdams Peil (Dutch vertical reference datum)
OMT	overall model test
PRF	Pulse Repetition Frequency
PS	Persistent Scatterer
PS1C	Persistent Scatterer candidate 1st order PS network
PS1	Accepted Persistent Scatterer 1st order PS network
PS2C	Persistent Scatterer candidate 2nd order PS network
PS2	Accepted Persistent Scatterer 2nd order PS network
PSI	Persistent Scatterer InSAR
RADAR	Radio detection and ranging
RD	Stelsel van de Rijksdriehoeksmeting (Dutch coordinate system)
RSR	Range Sampling Rate
SAR	Synthetic Aperture Radar
SCR	Signal to Clutter Ratio
SLC	Single Look Complex
SRTM	Shuttle Radar Topography Mission
WGS84	World Geodetic System 1984

cm	centimeter
km	kilometer
m	meter
mm	millimeter
yr	year

List of Symbols

A	design matrix
a	phase ambiguity
$B_\perp$	perpendicular baseline
c_m	compaction coefficient
D	depth of burial of a nucleus-of-strain
D_a	normalized amplitude dispersion
D_{ij}	displacement between PS i and PS j
$\underline{\hat{e}}$	vector of least-squares residuals
f_{dc}	Doppler centroid frequency
H	reservoir thickness
H_{ij}	(residual) topographic height between PS i and PS j
$\underline{n}$	measurement noise
p	reservoir pressure
Q_k	cofactor matrix for variance component estimation
Q_y	variance-covariance matrix of the observations
r	radial distance from the vertical axis through the nucleus-of-strain
SCR	Signal-to-Clutter ratio
$\underline{s}$	model imperfections
T	temporal baseline
T_q	teststatistic with q degrees of freedom
u_r	horizontal (radial) displacement at ground level
u_z	vertical displacement at ground level
V	volume of a nucleus-of-strain
v_{ij}	displacement rate between PS i and PS j
v_{sat}	satellite velocity
W	matrix that constructs double-difference observations
x	vector of unknown parameters
$\underline{y}$	vector of observations
ξ_{ij}	sub-pixel position in azimuth direction between PS i and PS j
η_{ij}	slant-range sub-pixel position between PS i and PS j
η,ξ	range and azimuth radar coordinates
ε_z	vertical strain
ν	Poisson's ratio
γ	phase coherence (in time)
γ^m	stack coherence for master m
$\underline{\hat{\sigma}}$	variance component estimator
θ	incidence angle
$\underline{\psi}$	phase observation in a single SAR scene
$\underline{\varphi}$	phase observation (wrapped or unwrapped is indicated)
$\underline{\varphi}_{ij}^k$	double-difference phase observation for the kth interferometric combination

Chapter 1
Introduction

Since the 1960s, various gas and oil reservoirs in the Netherlands have been taken into production. The largest is the Groningen gas reservoir with a thickness of 100–200 meters and a diameter of approximately 30 kilometers (NAM, 2005). Since the start of gas extraction, the reservoir layers have been compacting, which has resulted in subsidence at ground level up to 24.5 cm in 2003 (Schoustra, 2004). In the Netherlands, measuring subsidence due to gas and oil extraction is a legal obligation in order to take environmental counter measures when required. The environmental impact of hydrocarbon production has been stressed at the start of gas production in February 2007 from the subsurface below the Waddenzee, a protected marine wetland area (NAM, 2006). Here, the condition of near real time subsidence monitoring ('hand-on-the-tap') has been imposed, to avoid negative effects on the ecological system. These developments emphasize the need of geodetic subsidence monitoring techniques that regularly provide geodetic observations including their uncertainty boundaries.

1.1 Background

Ground level movements in the Netherlands have been measured periodically by means of leveling campaigns (de Heus et al., 1994; Schoustra, 2004). Measured height differences from precise leveling have a precision better than 1 $\text{mm}/\sqrt{\text{km}}$ (de Bruijne et al., 2005). Because of the long history of the leveling technique, the error budget is well known. Moreover, leveling networks are generally designed to incorporate redundant observations, that enable the testing and removal of erroneous measurements.

Although leveling is a well-established technique for subsidence monitoring, it has certain drawbacks: it is labor intensive and expensive. Moreover, it poses safety risks on subsidence monitoring, since measurements have to be taken along busy roads. Hence, the feasibility of spaceborne subsidence monitoring has been investigated since the development of satellite radar interferometry (InSAR) as a measurement technique from the 1990s. Various radar satellites have been operational, e.g., ESA's European Remote Sensing Satellites (ERS-1 and ERS-2), acquiring SAR imagery over an area of 100×100 km with a 35 days interval. The high temporal and spatial observation frequency of InSAR compared to leveling could potentially provide even additional insight in deformation mechanisms.

InSAR utilizes phase difference observations between two radar acquisitions for the estimation of surface deformation. Besides the deformation signal of interest, the interferometric phase also contains contributions due to atmospheric signal delay, (residual) topography and orbital errors (Hanssen, 2001). Moreover, only the

V.B.H. (Gini) Ketelaar, *Satellite Radar Interferometry*,
Remote Sensing and Digital Image Processing 14,

fractional phase is observed (the ‘wrapped’ phase), which implies that the number of integer cycles from satellite to the surface is unknown. The InSAR methodology that is aiming for a continuous coherent phase difference image (‘interferogram’) is referred to as conventional InSAR. Examples of the application of conventional InSAR for deformation monitoring are the displacement field of the Landers earthquake, demonstrated by Massonnet et al. (1993), and the detection of subsidence due to hydrocarbon production at the Belridge and Lost Hills oil fields in California, U.S.A., see e.g., van der Kooij (1997). Conventional InSAR is suitable for monitoring distinct deformation signals with a high magnitude compared to the error sources, in areas that do not suffer from temporal surface changes. However, in areas with low subsidence rates—subsidence due to gas extraction in the Netherlands is in the order of several mm/year—the estimation of error sources such as atmospheric disturbances is essential to obtain accurate deformation estimates. Furthermore, the loss of coherence due to temporal decorrelation of agricultural and vegetated areas limits the application of conventional interferometry.

To overcome the limitations of conventional interferometry, Persistent Scatterer (PS) interferometry has been introduced (Ferretti et al., 2000, 2001). Persistent Scatterers are targets with a coherent phase behavior in time, that often correspond with man-made features in the terrain. The phase difference observations from Persistent Scatterers are utilized for a dedicated estimation of the deformation signal and other phase contributions, such as topographic height differences and atmospheric disturbances. Moreover, they form a network of reliable measurement points in which phase unwrapping is performed. Persistent Scatterer InSAR (PSI) has been successfully applied in urban(ized) areas, where the PS density is high. PSI has been utilized for the estimation of deformation due to different causes: subsidence due to water pumping, mining activities, hydrocarbon production and landslides, see e.g., Fruneau (2003), Colesanti et al. (2005), Ketelaar et al. (2005), and Meisina et al. (2006). A quality assessment of Persistent Scatterer InSAR (PSI) has been performed by Colesanti et al. (2003): a precision of 1–3 mm for displacement estimates and 0.1–0.5 mm/year for deformation rates has been claimed (1-sigma).

1.2 Research Objectives

Because of the advances in the InSAR technique, potential applications have moved towards areas that are severely affected by temporal decorrelation, with low deformation rates over a large spatial extent. Subsidence due to gas extraction from the Groningen gas field is characterized by subsidence rates smaller than 1 cm/year that are distributed over a bowl shaped area with a diameter of more than 30 km. Furthermore, the area has a rural and agricultural character and is subject to varying atmospheric circumstances (Hanssen et al., 1999). Since the majority of the Groningen area suffers from temporal decorrelation, conventional InSAR is not applicable for subsidence monitoring. After the ERS revisit time of 35 days, only the urban(ized) areas appear coherent in the interferograms. Due to the small magnitude of the deformation signal of interest, the estimation of error sources and a precision and reliability assessment of the deformation estimates are essential. Preceding research on the applicability of InSAR in the Groningen area has been performed by Hanssen and Usai (1997) and Usai (2001). To overcome the effect of temporal decorrelation, the study focuses on coherent features (buildings, roads) in a selected number of interferograms with short baselines. The phase observations in several urban areas have been analyzed on their performance for subsidence estimation. Deviations from the existing deformation profiles have indicated that a more rigorous approach is required.

Although the quantification of the precision and reliability of PSI as a measurement technique is essential for subsidence monitoring, it is not sufficient for the estimation of the deformation signal of interest: subsidence due to hydrocarbon production. The radar satellite observes all kinds of surface deformation from space, regardless of the driving mechanism. A PSI displacement can represent both the instability of a building with a bad foundation, as well as shallow compaction due to ground water extraction or due to compacting soft soils in the shallow subsurface. Hence, both the precision and reliability of PSI as a measurement technique and the possible deformation causes have to be addressed. Moreover, the integration and cross-validation with historical leveling measurements have to be demonstrated for the operational use of PSI to ensure consistency in subsidence monitoring in the Netherlands.

Despite these complicating factors, the spatial and temporal observation frequency of PSI has potential to improve subsidence monitoring. For example, Odijk et al. (2003) show that the integration of InSAR and leveling observations results in an improved precision of the estimated subsidence parameters in areas that are poorly covered with leveling data. Moreover, the spatial and temporal observation frequency of PSI can potentially improve insight in reservoir behavior, and hence optimize the development of hydrocarbon reservoirs.

The identification of the main limitations and potential value of InSAR for subsidence monitoring in the Groningen area has lead to the following central research question:

Is the InSAR technique able to provide precise and reliable deformation estimates for the monitoring of subsidence due to hydrocarbon production in the Netherlands, particularly in the Groningen region?

This problem statement is divided into the following sub-questions:

1. Does the area of interest contain sufficient radar targets with coherent phase observations?
2. Does InSAR provide a precise estimation of surface displacements in the Groningen area?
3. How can we assess the reliability of InSAR deformation estimates?
4. Is it possible to estimate subsidence due to hydrocarbon production from InSAR measurements in the presence of multiple deformation phenomena?
5. Are the PSI deformation estimates in agreement with the leveling results?
6. Can InSAR aid the understanding of reservoir behavior?

7. Is the continuity of subsidence monitoring using InSAR guaranteed?

The first sub-question addresses the existence of Persistent Scatterers, which mainly correspond with man-made features in the terrain. From these targets, a network of double-difference phase observations (both spatial and temporal) is formed, enabling a dedicated estimation of the deformation signal of interest by means of geodetic adjustment and testing procedures. In urban areas, this technique has proven to be successful. However, the performance in highly decorrelated rural areas needs to be investigated.

The quality description of PSI estimates is divided into precision and reliability. Precision is the dispersion of a stochastic variable around its expectation value, whereas reliability expresses the detectability of model imperfections and their influence on the (deformation) parameter estimates. A complicating factor in the reliability assessment of PSI deformation estimates is the lack of redundancy in the estimation procedure when phase unwrapping cannot be assumed correct. To overcome this limitation, we introduce a reliability assessment that utilizes the PSI estimates from multiple independent tracks.

Besides the accuracy of PSI for the estimation of ground movements in the Groningen area, the *interpretation* of the deformation estimates is addressed. Since radar satellites monitor from space, they observe each surface movement, not only the signal of interest. A PS displacement may be caused by (a superposition of) several deformation causes: structural instabilities, shallow subsurface compaction, or hydrocarbon production. For the separation of the displacement components, knowledge on the physical nature of the PS is required: does the PS displacement represent a direct reflection from a well-founded building on a deep subsurface layer, or does it represent a multi-bounce reflection with the direct surroundings that are potentially affected by shallow compaction (Perissin, 2006)?

Cross-validation will be performed by a comparison with the displacement estimates from the leveling technique. The stochastics of both techniques are taken into account for the estimation of the deformation parameters in an integrated way. Since InSAR measurements can potentially replace (part of) the leveling measurements, future continuity should be guaranteed. Since the 'lifetime' of subsidence exceeds the lifetime of a radar satellite mission, this implies that the deformation estimates of multiple sensors have to be integrated.

Besides the scientific contribution, this research proposes an operational alternative for expensive and laborious leveling campaigns. The advantages of subsidence monitoring by remote sensing techniques using regular satellite acquisitions are obvious: it would significantly decrease the expenses for subsidence monitoring, and reduces the safety risks. Furthermore, both the temporal and spatial sampling frequency can be significantly increased: up to 4 acquisitions each 35 days in stead of one leveling campaign each 2–5 years, and more than 100 targets per km^2 in urban areas versus 1–2 benchmarks per km^2. Hence, besides a more controlled monitoring of subsidence due to hydrocarbon production, InSAR can potentially contribute to an increased insight in reservoir behavior.

1.3 Outline

Chapter 2 starts with a discussion on the formation of hydrocarbon reservoirs. Based on reservoir properties such as compaction coefficient and thickness, methods to predict subsidence at ground level are explained. Subsequently, the leveling campaigns in the Groningen area are addressed, followed by a review of the existing geodetic methodologies for the estimation of subsidence due to hydrocarbon production.

The PSI estimation theory and quality description is addressed in Chap. 3. It provides an overview of interferometric processing, PS selection, and the mathematical model of PSI. Subsequently, Chap. 4 focuses on the quality assessment of PSI. This comprises both the accuracy of the deformation estimates and the estimation of the deformation signal of interest in the presence of other deformation mechanisms ('idealization precision'). The lack of redundancy in the mathematical model that is introduced in Chap. 3 limits the reliability assessment of PSI. Hence, Chap. 5 introduces an alternative reliability assessment that utilizes PSI estimates from multiple independent satellite tracks that all observe the same deformation signal. The PSI estimates of these tracks are integrated through the so-called multi-track datum connection procedure.

The theoretical framework that is described in the Chaps. 3, 4, and 5 is applied for subsidence monitoring in the Groningen area in Chap. 6. The multi-track datum connection procedure that uses six ERS tracks further expands the monitoring area to the entire northeastern part of the Netherlands and a part of Germany. The precision of the displacement estimates are addressed for both ERS (1992–2005) and Envisat (2003–2007), as well as the effect of model imperfections. Furthermore, the idealization precision of the estimation of subsidence due to hydrocarbon production is investigated using PS characterization tools and a-priori knowledge on the spatio-temporal behavior of the deformation signal of interest.

An important condition for the operational use of PSI is the agreement of PSI displacement estimates with the historical leveling results. A strict comparison in terms of precision and accuracy is not straightforward, since leveling and PSI observations are physically different. Chapter 7 analyzes the correlation of displacement estimates from neighboring PS and leveling benchmarks. Furthermore, it addresses the spatio-temporal sampling of both techniques with respect to the precision of the observations and the estimation of the deformation parameters of interest.

Prior to the formulation of the conclusions and recommendations in Chap. 9, Chap. 8 discusses the results that have been obtained in Chaps. 6 and 7, and the future outlook. It shows the potential of PSI to increase knowledge on reservoir behavior, e.g., subsidence and subsequent uplift over an underground gas storage can be clearly identified due to the temporal sampling of PSI.

Chapter 2
Subsidence Due to Hydrocarbon Production in the Netherlands

In this chapter, the mechanism of subsidence due to hydrocarbon production is described. Subsidence at ground level is caused by the compaction of the reservoir rock due to hydrocarbon extraction. The spatial and temporal development of subsidence is dependent on the production rate, the physical reservoir rock properties and the overlying subsurface layers. In the Netherlands, subsidence monitoring is legally obliged to control the water management and to avoid environmental damage. Moreover, it provides information on reservoir behavior and well performance, for example to control steam injection for the optimization of oil production.

Section 2.1 starts with a brief overview of the geological circumstances that are required for the existence of hydrocarbon reservoirs. Subsequently, the geophysical properties of gas and oil reservoirs in the Netherlands, in particular the Groningen gas field, are discussed in more detail. Based on the reservoir parameters, models that have been developed for subsidence prediction are described. The actual subsidence measurements in the Netherlands are described in Sect. 2.2. Section 2.3 gives an overview of the subsidence estimation methodologies that have been applied since the start of gas extraction from the Groningen field.

2.1 Geological Background

In this section, the existence and properties of hydrocarbon reservoirs are discussed, followed by an explanation of the Groningen gas reservoir.

2.1.1 Hydrocarbon Reservoirs

Hydrocarbon is formed from organic debris exposed to high temperature and pressure due to increasing overburden (overlying sediments) in time (Chapman, 1983; Landes, 1959; Rondeel et al., 1996). It is stored in reservoir rock, where void spaces can be filled with water, hydrocarbon liquids (oil) or hydrocarbon gas. The most common reservoir rocks are sandstones and carbonates. The hydrocarbon compo-

V.B.H. (Gini) Ketelaar, *Satellite Radar Interferometry*,
Remote Sensing and Digital Image Processing 14,

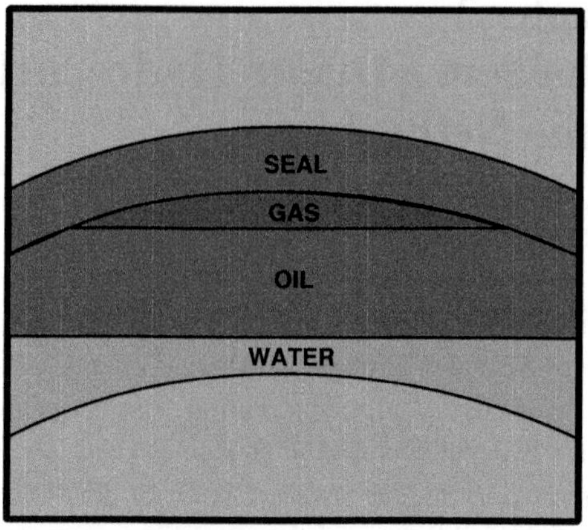

Fig. 2.1 Hydrocarbon accumulation in an anticlinal trap. The seal prevents the hydrocarbon fluids from migrating further upwards. Water, oil and gas are layered according to their density

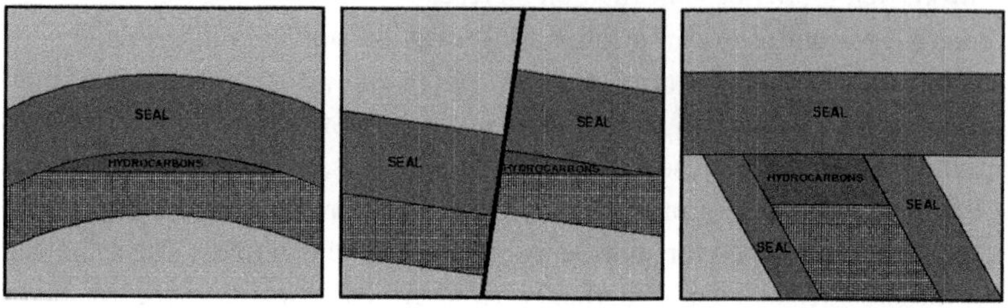

Fig. 2.2 Structural traps: anticlinal trap (left) and fault trap (middle). Stratigraphic trap (right): tilted layers of varying permeability

sition of a reservoir depends on the type of hydrocarbons, the temperature and the pressure in the reservoir. The fluids within a reservoir are layered according to their density, see Fig. 2.1.

Since the hydrocarbons tend to migrate upwards, a seal and a trap are required for hydrocarbons to accumulate. A seal consists of material that is impervious for hydrocarbon fluids. Examples of seals are shales or evaporites (such as salt layers). A trap is an enclosed reservoir that is that is surrounded by impervious rock. Traps are subdivided into structural and stratigraphic traps, see Fig. 2.2. Examples of structural traps are anticlines, faults and salt core structures. Stratigraphic traps are caused by changes in permeability. Tilting of sedimentary layers is often required for such a trap to exist (ibid.).

For a profitable extraction of hydrocarbons, a reservoir must meet certain quality criteria. Besides the hydrocarbon volume, the thickness and extent of the reservoir, porosity and permeability are driving factors. Porosity is the percentage of the total reservoir rock volume that is void space. Although porosity is required for hydrocarbon storage, it does not guarantee that the hydrocarbon fluids are able to flow in the reservoir. The ability of a rock to transmit fluid and discharge its hydrocarbon contents is defined as the permeability. The higher the permeability of a reservoir rock, the easier the hydrocarbon fluids will flow. Porosity and permeability are dependent on grain shape, packing and sorting, degree of cementation and the overburden. For more information, see Craft and Hawkins (1991) and Dake (2002).

2.1.2 The Groningen Reservoir

The subsurface of the Netherlands contains numerous gas fields and several oil reservoirs. The majority of the hydrocarbon reservoirs is situated in the northeastern part of the Netherlands, see Fig. 2.3. Oil and gas production in the Netherlands has started with the discovery of the Schoonebeek oil field in 1943 and the Coevorden gas field in 1948. The Groningen gas field was discovered in 1959.

The geology of the Groningen gas field is depicted in Fig. 2.4. The gas has been formed in the Carboniferous period (365–290 million years ago). Subsequently, it has migrated upwards to the porous sandstones in the Rotliegend layer from the Permian period (290–250 million years ago). These sandstone layers have been formed from aeolian and fluvial deposits (de Jager and Geluk, 2007). The aeolian deposits form the best reservoirs since the grains are well sorted. The gas reservoirs are sealed by the Ten Boer claystone layer and the thick Zechstein salt layer. The boundaries of the Groningen gas reservoir are mainly defined by fault zones, with a few closures that are caused by the orientation of the layers with respect to the horizontal plane (NAM, 2003c).

The Groningen gas field has a horizontal extent of approximately 900 km^2. It is situated at a depth of 2750–2900 m and its thickness varies between 100 and 200 m (NAM, 2005). Porosity values vary between 16 and 20% (Teeuw, 1973). The Groningen gas field is the largest gas field in western Europe and one of the largest gas fields in the world. The estimated recoverable volume is $\sim$2700 billion m^3. The total number of wells that has been established is 295, arranged in 29 clusters. Gas production has started in 1963. Currently, the focus lies primarily on the gas production from the smaller gas fields in the Netherlands (NAM, 2003c). The production from the Groningen gas field is kept relatively low ($\sim$30 billion m^3 per year) to increase its lifetime.

Since the focus of this book lies on subsidence monitoring, the reader is referred to Duin et al. (2006); Lutgert et al. (2005), and Breunese et al. (2005) for a detailed description of the geology of the Netherlands and the performance of gas production from the Groningen field.

2.1.3 Reservoir Properties

This section explains the reservoir properties that determine the potential amount of subsidence due to hydrocarbon production. First, the driving factors for the compaction of the reservoir are explained. Secondly, the influence of faults and aquifers are explained.

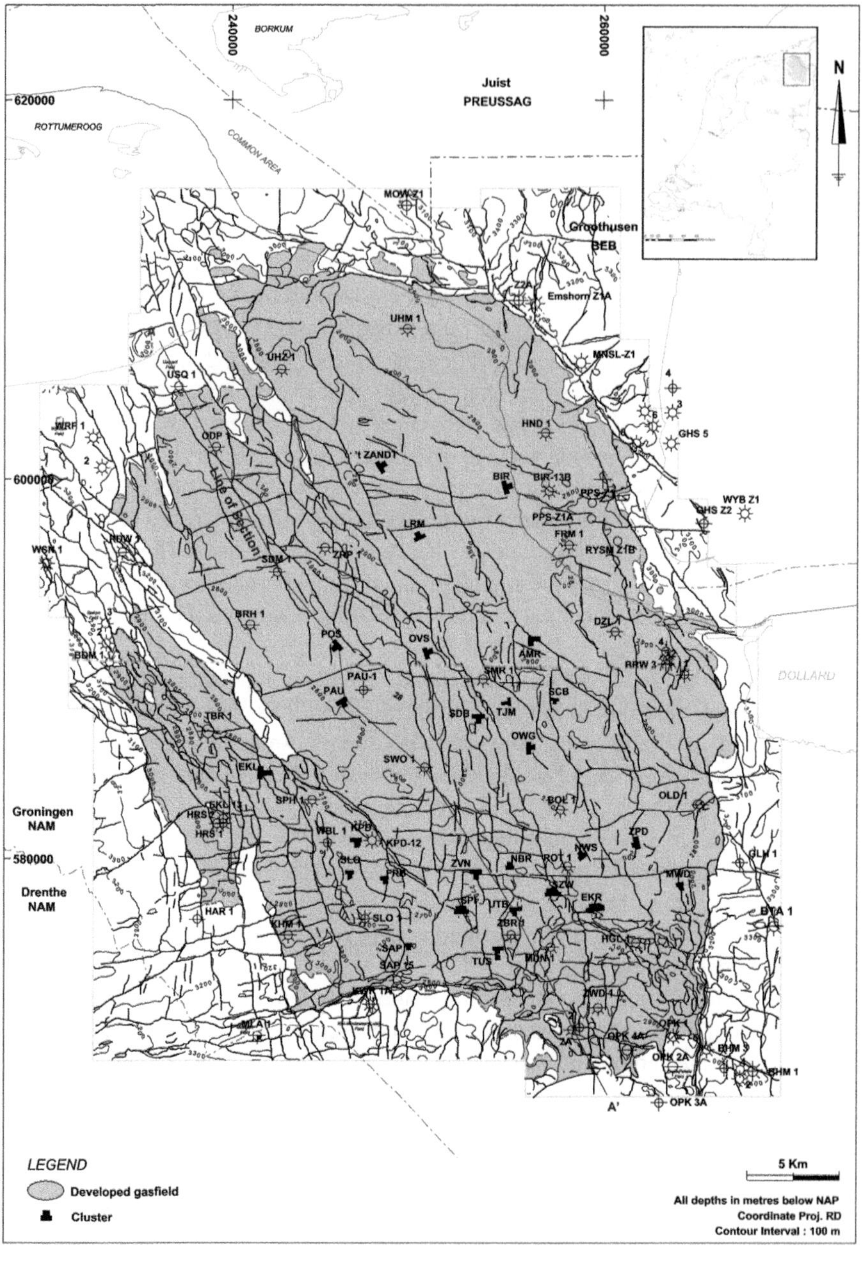

Fig. 2.3 Overview of the location and spatial extent of the Groningen gas field, including the well locations and the faults (NAM, 2003c)

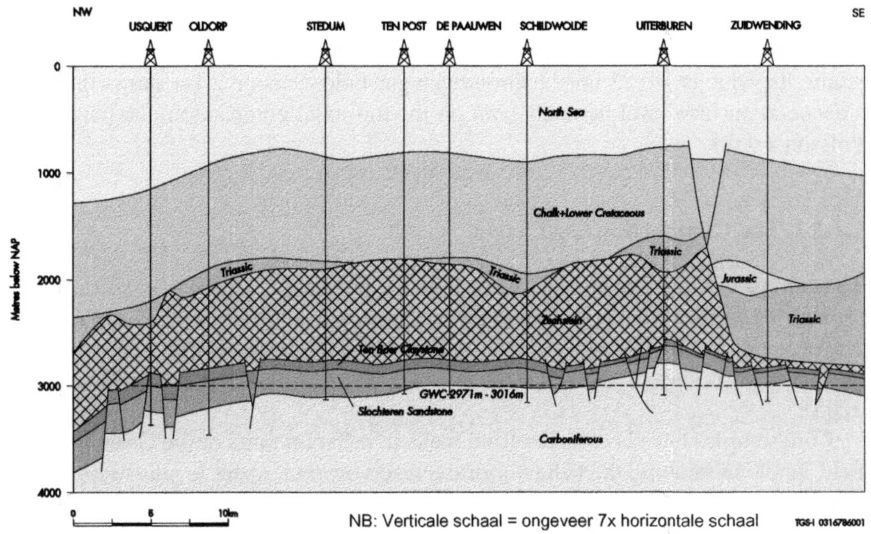

Fig. 2.4 Cross-section of the Groningen gas field (NAM, 2003c). The location of this cross-section is indicated in Fig. 2.3 as 'Line of Section'. The Slochteren sandstone formation is part of the Rotliegend layer

2.1.3.1 Reservoir Compaction

During the production of gas and oil, the pore pressure decreases. Because the overburden remains unchanged, the effective stress on the grain structure of the reservoir increases. As a result, the reservoir is compacting: its volume decreases. If the lateral dimension of the reservoir is large compared to its thickness, compaction mainly results in a reduction of reservoir height (Geertsma, 1973b). Hence, reservoir compaction can initially be characterized by the vertical strain ε_z in the reservoir:

$$\varepsilon_z = \frac{dz}{z}, \tag{2.1}$$

which is the change in reservoir height dz relative to its initial height z, caused by an increase in effective stress due to a decrease in pore pressure dp under a constant overburden. Reservoir compaction in vertical direction is characterized by the uni-axial compaction coefficient c_m:

$$c_m = -\frac{1}{z}\frac{dz}{dp}, \tag{2.2}$$

that describes the compaction per unit change in pore pressure (in bar^{-1}). The total compaction ΔH until a certain point in time is dependent on the difference in pore pressure Δp since the start of the production and the initial reservoir thickness H:

$$\Delta H = c_m \cdot \Delta p \cdot H. \tag{2.3}$$

The compressibility of the reservoir rock in lateral direction is specified by Poisson's ratio ν. Poisson's ratio is the ratio between the lateral strain and the vertical strain. Its value is ~0.25 for the Groningen gas field. Section 2.1.4 shows that subsidence at surface level depends both on the uni-axial compaction coefficient and Poisson's ratio.

2.1.3.2 The Compaction Coefficient

The compaction coefficient is dependent on the physical reservoir properties. There are two methods available to derive the compaction coefficient: by laboratory tests on core samples from the wells (Teeuw, 1973), and by means of radio-active bullets that have been shot in the reservoir at observation wells (de Loos, 1973; NAM, 2005).

Core samples have been taken from wells in different parts of the Groningen gas field. In the laboratory, the behavior of the reservoir rock under in situ stress conditions has been analyzed. The reservoir compaction is determined from the relative change in reservoir thickness due an increase in effective vertical stress under zero lateral strain.

Besides by performing laboratory tests on core samples, compaction can be measured in situ. The measurement targets for in situ compaction measurements are radio-active bullets that have been shot in the formation at regular distance. Their relative displacement is measured periodically by means of a gamma-ray detector. This gamma-ray detector is connected to a cable that is deployed in an observation well. In the Groningen gas field, eleven observation wells have been established where these in situ compaction measurements are carried out with millimeter precision (NAM, 2005). The compaction measurements show a linear dependency on the reservoir pressure. The compaction coefficient c_m that has been deduced from these measurements varies between 0.45 and $0.75 \cdot 10^{-5}$/bar (ibid.).

The initial pressure in the Groningen reservoir was 347 bar, which has dropped to 125 bar in 2005 (ibid.); the average thickness of the reservoir is 170 m. Using (2.3), this would imply that gas production up to 2005 has caused a total reservoir compaction between 17 and 28 cm. The resulting compaction at ground level is dependent on the depth and radius of the reservoir and Poisson's ratio as well, see Sect. 2.1.4. Furthermore, reservoir compaction may be subject to a delay in time (Hettema et al., 2002), in which the reservoir reconverges to an equilibrium and the compaction propagates through the overburden to ground level.

The amount of reservoir compaction is also driven by the reservoir rock properties: ordering, shape and hardness of the grains, and the degree of cementation or frame rigidity (Teeuw, 1973). The rock properties also determine whether the deformation is reversible or not. The deformation of hard rock exhibits in general elastic (reversible) behavior. The compaction of soft rock may be partly irreversible due to crushing and relocation of grains. Rock types are subdivided in tight rock, well-consolidated rock, semi-consolidated rock and unconsolidated rock, with varying porosity from 0 to 40%, and an increasing compaction coefficient. The Groningen Rotliegend reservoir is classified as semi-consolidated, and elastic behavior is assumed.

2.1.3.3 Reservoir Connectivity

The amount of reservoir compaction is dependent on the thickness of the reservoir, the pressure drop in the reservoir, and the compaction coefficient of the reservoir rock. If these parameters vary through a reservoir, the reservoir compaction will vary as well. Discontinuous changes can be found near faults. Depending on the reservoir thickness, the vertical offset, the orientation of the faults, and the depth of the gas-water contact, they can be sealing or not. If the drainage region of a well contains sealing faults, the well will not produce hydrocarbons from the disconnected block. As a result, there will a be compacting and non-compacting block on either side of the fault. The fault pattern of the Groningen gas field is depicted in Fig. 2.3. It has a dominantly southeast–northwest orientation. The interaction between the reservoir blocks can impose an uncertainty on the subsidence prediction.

2.1.3.4 Aquifers

In Sect. 2.1.1 it has been explained that a reservoir can be partly filled with water. The part of the reservoir that is filled with water is called the aquifer. Due to the higher density of the water, the aquifer will be located below the hydrocarbon fluids. The presence and dimensions of the aquifer determines the pressure drop during hydrocarbon production. If the aquifer is large with respect to the gas reservoir, it can provide pressure support to the hydrocarbon reservoir (NAM, 2005). If the aquifer is small, hydrocarbon production can significantly affect the aquifer pressure. Since the aquifer partly determines the pressure distribution within the reservoir, knowledge on the depletion of aquifers is important for the estimation of reservoir compaction. Moreover, uncertainties can exist about the connection of lateral aquifers around fault zones. Since there are hardly any wells established in the aquifer zones, there is a lack of observations on their pressure behavior. Geodetic measurements at ground level, such as leveling and PSI, can provide knowledge on the depletion of aquifers. For example, based on the leveling campaigns, it could be concluded that the aquifer to the west of the Groningen gas field is not depleting (ibid.).

2.1.4 Subsidence Prediction Methodologies

Based on the geophysical properties of the hydrocarbon reservoir and the overlying layers, subsidence at ground level can be predicted. Various methods have been applied: analytical (Geertsma, 1973a), semi-analytical (Fokker, 2002; Fokker and Orlic, 2006), numerical (Sroka and Hejmanowski, 2006) and finite element methods (Geertsma and van Opstal, 1973; Fredrich et al., 2000).

The analytical solutions for subsidence prediction that are described by Geertsma (1973a) assume that the overburden is uniform and elastic. The reservoir itself is built up of so-called 'nuclei-of-strain' that have a small but finite volume V. The vertical displacement u_z caused by a nucleus-of-strain is given by:

$$u_z(r,0) = -\frac{c_m(1-\nu)}{\pi}\frac{D}{(r^2+D^2)^{3/2}}\Delta p V, \tag{2.4}$$

where:

r	radial distance from the vertical axis through the nucleus-of-strain,
c_m	uni-axial compaction coefficient(kg/cm^2)$^{-1}$, see (2.2),
ν	Poisson's ratio,
Δp	pore pressure reduction (kg/cm^2),
D	depth of burial of the nucleus-of-strain,
V	volume of the nucleus-of-strain.

A negative vertical displacement implies subsidence, whereas a positive vertical displacement implies uplift. The geometrical shape of the displacement induced by a nucleus-of-strain is equal to the displacement induced by a point source as defined by Anderson (1936) and Mogi (1958).

Surface deformation due to hydrocarbon production is not restricted to vertical displacements. The horizontal displacement u_r due to a nucleus-of-strain reads:

$$u_r(r,0) = +\frac{c_m(1-\nu)}{\pi}\frac{r}{(r^2+D^2)^{3/2}}\Delta p V, \tag{2.5}$$

where a positive horizontal displacement is in the direction towards the location of the nucleus-of-strain. From (2.4) and (2.5) it is deduced that the ratio between horizontal and vertical displacements equals $-r/D$.

The total subsidence above a reservoir can subsequently be obtained by the integration of the nucleus-of-strain solutions over the entire reservoir. A closed form solution of the integration of the nucleus-of-strain solutions is given by Geertsma (1973a), based on a simplified representation of the reservoir as a disc-shaped reservoir of thickness H and radius R at depth D. The pressure reduction Δp is assumed to be uniform through the reservoir. The equations are non-linear and require the evaluation of Hankel-Lipschitz integrals. The maximum vertical displacement above a disc-shaped reservoir can be expressed analytically:

$$u_z(0,0) = -2c_m(1-\nu)\Delta p H\left(1 - \frac{D/R}{\sqrt{1+(D/R)^2}}\right). \tag{2.6}$$

Apart from the compaction coefficient, Poisson's ratio, the pressure drop and the thickness of the reservoir, the ratio between the depth and the radius of the reservoir determines the maximum amount of subsidence.

The analytical expressions for subsidence prediction are based on a simplified representation of the subsurface. The reservoir is not a perfect disc; nor is the overburden perfectly homogeneous. Hejmanowski and Sroka (2000) subdivide the reservoir into elementary cubicoids with each their own geomechanical properties (thickness, compaction, pressure drop). Subsequently, influence functions are applied to estimate subsidence due to a reservoir element at surface level. The total subsidence is the superposition of the contributions of all reservoir elements. Finite element methods utilize a geomechanical model of the entire subsurface: the reservoir and the adjacent geological layers, up to ground level. Fredrich et al. (2000) model the evolution of the displacements in the Belridge reservoir and the overburden, based on such a finite element model.

Finite element models have the advantage that they can be applied to reservoirs of arbitrary geometry with varying reservoir properties and pressure distribution, see e.g., Geertsma and van Opstal (1973). Hence, a more accurate prediction of vertical displacements and horizontal gradients can be obtained, provided that the distribution of deformation properties in the reservoir is known sufficiently. Moreover, the overburden can be modeled more accurately in finite element models. A disadvantage of finite element methods is the computation time. Hence, semi-analytical modeling (Fokker, 2002; Fokker and Orlic, 2006) has been introduced. Semi-analytical modeling avoids the time consuming finite element approach, but uses a more sophisticated model for the subsurface compared to the analytical solutions of Geertsma (1973a). Instead of assuming a homogeneous subsurface, the subsurface is divided into multiple layers with each their own (visco-)elastic properties.

Both the analytical method from Geertsma (1973a) (with the reservoir subdivided into smaller blocks) and finite element analysis have been applied to predict subsidence due to hydrocarbon production in the Netherlands (NAM, 2005). Since the results of both methods are comparable, the analytical method is used for the majority of the gas fields (ibid.). Finite element analysis has been applied to calculate subsidence above the Ameland reservoir: it is located below a complicated salt structure which behavior cannot be modeled using the analytical method (ibid.).

2.2 Subsidence Monitoring Using Leveling Measurements

This section addresses the leveling campaigns that have been performed for subsidence monitoring in the Netherlands, and the legal guidelines.

2.2.1 Leveling Campaigns

Since the start of gas production in Groningen in the 1960s, leveling campaigns have been performed periodically. Leveling is an optical land surveying technique that measures height differences between established benchmarks. These benchmarks are spatially distributed over the area of interest and are ideally a discretization of the shape of the subsidence pattern. By measuring the benchmark height differences in multiple epochs, the development of the subsidence bowl(s) is monitored.

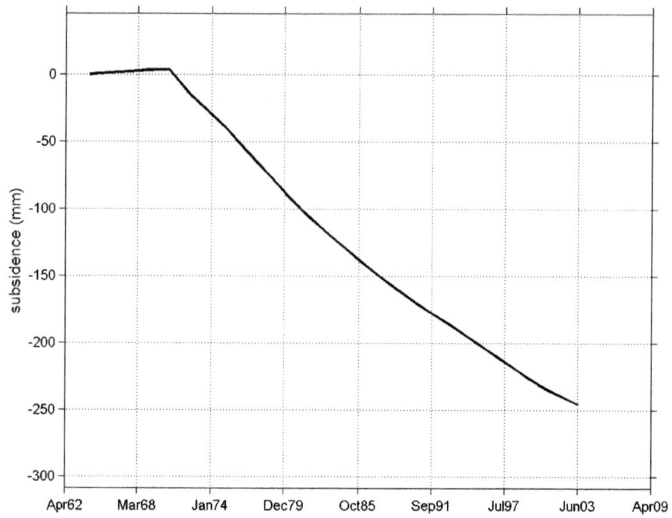

Fig. 2.5 Subsidence (mm) since the start of gas production in the center of the Groningen subsidence bowl, estimated from leveling observations (Schoustra, 2004). The subsidence rate has slightly decreased after the change of priority to the smaller gas fields from the 1970s

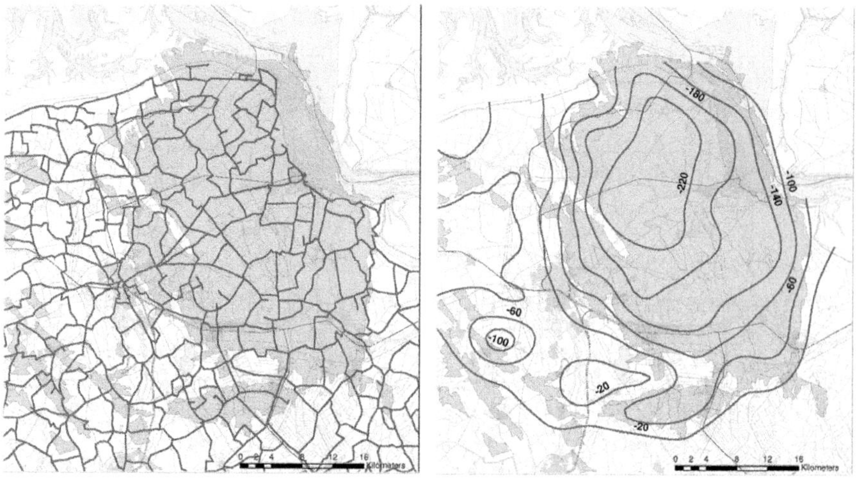

Fig. 2.6 Leveling network of the 2003 campaign (left) and subsidence (mm) since the start of gas production up to 2003 (right). The gas fields are depicted in green. The total length of the leveling trajectories is ~1000 km. Source topographic background: © Kadaster en de Openbare Registers, Apeldoorn

Since subsidence measurements are decisive for taking environmental countermeasures, the quality assessment of the estimated height differences is crucial. Redundant measurements have been taken in order to test observations on measurement errors and systematic errors. Figure 2.6 shows the leveling loops in the northeastern part of the Netherlands. Subsidence due to gas extraction has been estimated from repeated leveling campaigns since the start of the production. The deepest point in the Groningen subsidence bowl has subsided 24.5 cm until 2003 (Schoustra, 2004), see Fig. 2.5. In time, the displacements are approximately linear. After the 1970s the displacement rate has slightly decreased, after the priority has been shifted to the smaller gas fields. The Groningen gas field and the underground gas storages in Norg and Grijpskerk have the role of swing producer, covering peaks in demand (NAM, 2008).

Benchmark heights are orthometric and refer to the local Dutch height reference system 'Normaal Amsterdams Peil' (NAP). Since leveling is a relative technique, all heights are estimated with respect to a reference benchmark.

2.2.2 Legal Guidelines

Monitoring surface deformation due to mineral extraction is legally obliged in the Netherlands according to the Dutch mining legislation (Mijnbouwwet, 2008). Oil, gas and salt mining companies are obliged to develop and update measurement plans that need to be approved by the Ministry of Economic Affairs. According to the mining legislation, these measurement plans should contain the points in time, the location of the measurements, and the measurement techniques. One measurement campaign needs to be performed prior to the start of the production. The State Supervision of Mines (the mining authority) supervises all mining activities in the Netherlands, including the estimation of subsidence due to mineral extraction. Its mission is to ensure that the exploration and production of minerals in the Netherlands and the Dutch part of the continental shelf is carried out in a responsible and socially acceptable manner (SodM, 2008).

Duquesnoy (2002) defines further guidelines for subsidence monitoring using the leveling technique. A condition that is imposed on the leveling measurements is their agreement with the precision criteria as defined by AGI (2005). Examples of these precision criteria are the critical values for the misclosures of leveling loops and the precision of the height difference observations.

Furthermore, Duquesnoy (2002) investigates the spatial and temporal observation density. The required spatial benchmark density is dependent on the shape and extent of the subsidence bowl. Guidelines are provided based on a simplified representation of the gas reservoir. The Groningen gas field may be approximated by a disc-shaped reservoir of radius 15 km at a depth of 3 km. Based on an angle of draw of 45 degrees, the subsidence border is located at 18 km from the center of the reservoir. Application of the guidelines of Duquesnoy (2002) would imply a benchmark density of 1 per km^2 in the deepest part of the bowl and outside the subsidence border. At the slopes, a slightly higher spatial density (1.5 benchmark per km^2) is required to reconstruct the spatial subsidence pattern.

The determination of the measurement frequency is dependent on the precision of subsidence measurements. A new measurement campaign will only contribute if the expected subsidence is significant compared to the measurement precision. Moreover, the subsidence history based on preceding measurements can be utilized to predict subsidence with a higher precision. In this way, the measurement frequency can be lowered as the monitoring period increases, provided that the production rate does not change significantly and the compaction process is linear elastic. However, for practical reasons, a maximum period of 5 years is maintained between two measurement campaigns (ibid.).

2.3 Geodetic Deformation Monitoring

Geodetic techniques for deformation monitoring have been refined through the years. From the estimation of benchmark height differences, deformation monitoring has been extended with the parameterization of the temporal and spatial behavior of the deformation phenomenon. Moreover, dynamic systems that describe the forces and loads that cause the deformation and the physical properties of the deformation mechanism can be included, see e.g., Welsch and Heunecke (2001). Hence, advanced deformation analysis requires an interdisciplinary approach that integrates geodetic and geophysical skills.

This section provides an overview of deformation monitoring methodologies that have been applied to estimate subsidence due to gas extraction in the Netherlands. It starts with an overview of geodetic adjustment and testing techniques. Subsequently, point-wise multi-epoch deformation analysis is explained. This type of analysis is followed by the estimation of continuous spatio-temporal deformation phenomena.

2.3.1 Adjustment and Testing Procedure

The adjustment and testing procedure underlying geodetic deformation monitoring performs the estimation of unknown parameters and testing on observational and model errors in an integrated way. The testing procedure is important in deformation analysis, since the optimal parameterization of the unknowns is often less well known compared to classical geodetic applications, such as cadastral surveys. For example: the spatial shape of subsidence due to gas extraction has a higher degree of uncertainty than the location of the corner of a house. Hence, multiple alternative hypotheses are evaluated to determine the mathematical model that minimizes the least-squares residuals with respect to the observations. In this way, the optimal deformation model is found for the signal of interest. Of course each of the models under the alternative hypotheses should be physically explainable, to avoid fitting a non-realistic model to the observations. This section summarizes the mathematical framework of the adjustment and testing procedure.

The system of equations under the null hypothesis H_0 is formulated as a Gauss-Markov model:

$$H_0: \quad E\{\underline{y}\} = Ax; \quad D\{\underline{y}\} = Q_y, \tag{2.7}$$

where $\underline{y}$ is the vector of observations, x are the unknown parameters, and the design matrix A specifies the functional relation between them. The underlining of a vector (such as $\underline{y}$) indicates its stochastic character; the variance-covariance matrix of the observations is represented by Q_y.

Estimates of the unknown parameters are obtained by least-squares adjustment (Teunissen, 2000a). Subsequently, the validity of the null hypothesis is tested in the Detection, Identification and Adaptation (DIA) procedure (Teunissen, 2000b). In the **Detection** step, the null hypothesis is tested by means of the overall model test (OMT):

$$\underline{T}_{q=m-n} = \underline{\hat{e}}^T Q_y^{-1} \underline{\hat{e}}; \qquad \text{reject } H_0 \text{ if } T_{q=m-n} > \chi^2_\alpha(m-n, 0), \tag{2.8}$$

which is dependent on the least-squares residuals $\underline{\hat{e}}$ and the variance-covariance matrix of the observations. The redundancy $m-n$ equals the number of observations minus the number of unknowns (provided that the design matrix is of full rank).

If the overall model test is rejected, alternative hypotheses can be specified that are evaluated in the **Identification** step:

$$H_a: \qquad E\{\underline{y}\} = Ax + c_y \nabla, \tag{2.9}$$

where ∇ represents the model error and c_y specifies the functional relation with the observations that can be multi-dimensional. A standard test is datasnooping, where individual observations are checked for blunders. In this case, c_y will have the shape

$$c_{y_i} = (0, \ldots, 0, 1, 0, \ldots, 0)^T \tag{2.10}$$

for the ith observation.

The functional model that parameterizes the signal of interest is often not very well known a-priori in deformation monitoring. Therefore, multiple tests of different dimensions are specified to trace different kinds of model deviations. In the evaluation of tests of different dimensions, the one with the lowest teststatistic does not necessarily correspond with the most likely alternative hypothesis. This is caused by the different probability density functions for tests of different dimensions. A solution is provided by de Heus et al. (1994) by introducing testquotients: the ratio of teststatistics and their critical values. Provided that the power of the test is set to 50%, testquotients can be directly compared.

The **Adaptation** step involves either remeasuring and replacing (a part of) the observations or the replacement of the null hypothesis by the most likely alternative hypothesis. To test the validity of the mathematical model after adaptation, the DIA procedure is performed in an iterative way.

Besides the functional model, the stochastic model can be re-evaluated as well by means of variance component estimation (VCE) (Teunissen, 1988; Amiri-Simkooei, 2007). The stochastic model is then decomposed for the estimation of the variance factors σ_k^2:

$$Q_y = \sum_{k=1}^{p} \sigma_k^2 Q_k, \tag{2.11}$$

where Q_k are the cofactor matrices. The estimates for the variance components $\underline{\hat{\sigma}}$ are obtained by solving the following system of equations:

$$\underline{\hat{\sigma}} = N^{-1} l, \tag{2.12}$$

where:

$$N_{kl} = \text{tr}(Q_y^{-1} P_A^{\perp} Q_k P_A^{\perp} Q_l); \qquad l_k = \underline{\hat{e}}^T Q_y^{-1} Q_k Q_y^{-1} \underline{\hat{e}}, \tag{2.13}$$

where k and l are the row and column index for the kth and lth variance factor. The required input for VCE stems from the mathematical model and the adjustment results:

- $\underline{\hat{\sigma}}$ the estimator for the variance components σ_k^2,
- $P_A^{\perp}$ the orthogonal projector: $P_A^{\perp} = I - A(A^T Q_y^{-1} A)^{-1} A^T Q_y^{-1}$, and
- $\underline{\hat{e}}$ vector of least squares residuals: $\underline{\hat{e}} = P_A^{\perp} \underline{y}$.

Since Q_y itself is involved in VCE, the variance factor estimates are obtained in an iterative way. The precision of the variance component estimates follows from the propagation law:

$$Q_{\hat{\sigma}} = N^{-1}. \tag{2.14}$$

Verhoef et al. (1996) describe the Detection-Identification-Adaptation procedure for deformation analysis incorporating VCE. Besides the estimation of variance factors (for example to estimate the measurement precision), other stochastic parameters such as the spatial correlation length of the residual signal can be obtained as well through VCE. The decomposition of Q_y in such a situation is explained in Sect. 4.3.

2.3.2 *Point-wise Multi-epoch Deformation Analysis*

In point-wise multi-epoch deformation analysis, the deformation signal of interest is represented by discrete measurement points that are monitored at subsequent points in time. An example is 1D deformation analysis using leveling measurements from multiple epochs (de Heus et al., 1994). It can be subdivided into the following steps:

1. epoch analysis: free network adjustment and testing of leveling height difference observations per epoch,
2. stability analysis: stability testing of underground benchmarks that are located outside the subsiding area,
3. deformation parameter estimation.

The last step restricts to the temporal analysis of benchmark height estimates, in a *static* or *kinematic* way. In static deformation analysis, subsidence per benchmark is computed by subtracting the estimated height from the initial height, whereas kinematic deformation analysis models the displacements in time: a polynomial is fit through the height estimates, or an estimation is performed of geophysical parameters that are driving factors for the observed deformation.

Kinematic deformation analysis can be further subdivided into a *deterministic* and a *stochastic* approach. The deterministic approach attributes all residuals to measurement noise. The stochastic approach includes a residual component that can be addressed to model imperfections due to the simplification of the actual deformation pattern.

2.3.3 Continuous Spatio-temporal Deformation Analysis

Point-wise multi-epoch deformation analysis can be further developed to deformation analysis in which the continuous spatio-temporal evolution of the deformation phenomenon is modeled. Depending on the a-priori knowledge on the deformation signal of interest, this requires not only functional modeling, but also stochastic modeling of model imperfections. This section illustrates the application of continuous spatio-temporal deformation analysis for subsidence phenomena.

2.3.3.1 Subsidence—Functional Model

The spatial evolution of subsidence over multiple epochs can be described by, e.g., a point source model, or an ellipsoidal model, or prognosis grids, e.g., based on geomechanical modeling of the reservoir and the subsurface.

Depending on the complexity of the deformation mechanism, subsidence can optionally be estimated as a superposition of point source or ellipsoidal models.

The point source concept stems from volcanic applications. These point sources are often referred to as Mogi sources (Anderson, 1936; Mogi, 1958). Okada (1992) derives the displacement field in a homogeneous half-space due to point sources of

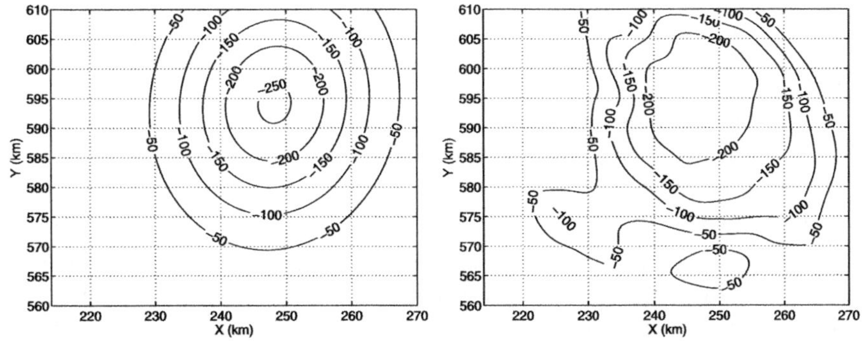

Fig. 2.7 Subsidence (mm) above the Groningen gas field estimated as a mathematical ellipsoidal shape (left) and subsidence prognosis based on geomechanical modeling of the reservoir and the subsurface (right)

different types. Specific directional point source types are defined for displacement fields due to earthquakes. If only vertical displacement due to a single point source is considered, its estimates at surface level read:

$$u_z(r,0) = M\frac{D}{(r^2 + D^2)^{3/2}}, \tag{2.15}$$

where M is a multiplication factor. The physical parameters involved in this multiplication factor are dependent on the application: deformation due to gas extraction, earthquakes or volcanic activities. It can be a function of the forces acting on the deforming body, shear modulus, compaction coefficient, Poisson's ratio, pressure change, and volume changes.

Note that (2.15) is similar to (2.4). The geometrical shape of the point source model is equal to the analytical expression for the vertical displacement due to a nucleus-of-strain. The multiplication factor from (2.15) consists of the compaction coefficient, Poisson's ratio, the pressure drop and the volume of the nucleus-of-strain. The analogy of the integration over the reservoir shape (Sect. 2.1.4) would be the superposition of the contribution of multiple point sources.

The second parameterization type models subsidence as (a superposition of) ellipsoidal bowls (Kenselaar and Quadvlieg, 2001). In time, the displacement rate is assumed linear. The subsidence velocity decreases exponentially with the distance to the center of the bowl. The subsidence z at time t for point i on a certain location reads:

$$z_i^{t_0 t} = \begin{cases} \dot{z}(t-t_0)e^{-\frac{1}{2}r_i^2} & \text{for } t \geq t_0 \\ 0 & \text{for } t < t_0 \end{cases} \tag{2.16}$$

with:

$$r_i = \sqrt{\frac{((x_i - x_c)\sin\phi + (y_i - y_c)\cos\phi)^2}{a^2} + \frac{((x_i - x_c)\cos\phi - (y_i - y_c)\sin\phi)^2}{b^2}}, \tag{2.17}$$

where:

$\dot{z}$	displacement rate of point i,
r_i	distance of point i to the center of the subsidence bowl,
t_0	starting time of subsidence,
x_i, y_i	location of point i,
x_c, y_c	location of the center of the bowl,
ϕ	orientation of the bowl,
a, b	length of the ellipsoidal axes.

The last model type is the subsidence prognosis: based on geophysical reservoir behavior and the overburden, displacements are estimated in the area of interest in a regular grid that is subdivided in blocks (the grid cells). The spatial variation of geophysical parameters is taken into account in the subsidence prognosis. Hence, the subsidence prognosis is more likely to provide a realistic subsidence prediction, compared to the point source and the ellipsoidal model. The point source and the ellipsoidal model tend to simplify the subsidence pattern. For all three model types holds that model deviations have to be assessed. If these model deviations can be explained by a geophysical mechanism, the functional model can be improved.

2.3.3.2 Subsidence—Stochastic Model

Since the functional deformation model is generally not well known a-priori, model imperfections are often incorporated in the stochastic model. This requires the model imperfections to be modeled by a covariance function that describes the spatio-temporal behavior. As a result, the variance-covariance matrix does not only contain the measurement noise component $\underline{n}$, but also the model imperfections $\underline{s}$:

$$Q_y = Q_{nn} + Q_{ss}. \tag{2.18}$$

Model imperfections comprise uncertainties in both the parameterization of the deformation signal itself and the physical representation of the measurement points. For example: if the measurement points exhibit additional autonomous movements due to shallow subsurface displacements, their displacements do not unambiguously represent subsidence due to gas and oil extraction. These autonomous movements can be stochastically modeled as spatially uncorrelated but temporally correlated movements. An example of autonomous movements are settlement movements of benchmarks that are used in a leveling network, see for example Sect. 6.5.4. A well-known settlement model is the Koppejan model (Verruijt and van Baars, 2005), which models the settlement as a logarithmic function of time. Since $\lim_{t\to\infty}\log_{10}(t) = \infty$, settlement behavior as a model imperfection is unbounded and therefore the variogram and its corresponding covariance function do not exist. Autonomous movements can stochastically be modeled as a random-walk process (Odijk and Kenselaar, 2003) or by an empirical covariance function (Houtenbos, 2004):

$$\sigma^2_{s_i^t} = \sigma_s^2|t-t_0|^{2p}, \qquad \sigma_{s_i^t s_i^u} = \frac{1}{2}\sigma_s^2(|t-t_0|^{2p} - |t-u|^{2p} + |u-t_0|^{2p}), \tag{2.19}$$

where t_0 is the reference time before the start of the subsidence, t and u are points in time, and p is the power of the empirical covariance function. If $p = 0.5$, this empirical covariance function reduces to the random walk model, see e.g., Chatfield (1989). In the estimation of subsidence due to gas extraction from leveling measurements in Groningen for 2003, noise due to autonomous benchmark movements was set to 0.2 mm/$\sqrt{\text{yr}}$ (Schoustra, 2004).

Model imperfections due to uncertainties in the parameterization of the subsidence signal or its prognosis can be stochastically described by a covariance function as well. An example is the covariance function from Houtenbos (2004) that models spatially and temporally correlated deviations between measurements and the subsidence prognosis:

$$\sigma_{z_i^t z_i^u} = \frac{1}{2}\sigma_z^2(|t-t_0|^{2p} - |t-u|^{2p} + |u-t_0|^{2p})e^{-(l_{ij}/L)^2}, \tag{2.20}$$

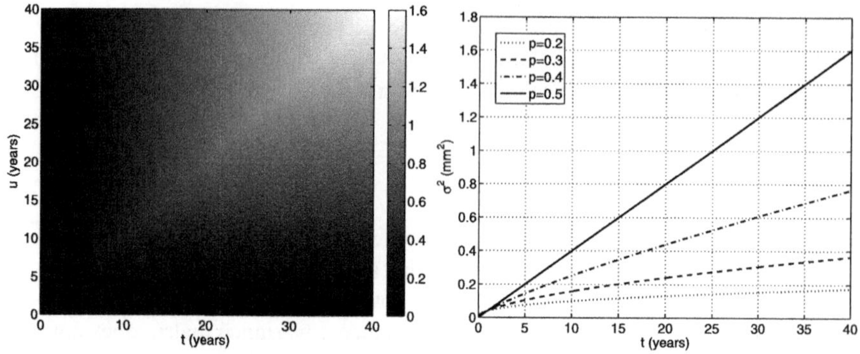

Fig. 2.8 Left: covariance induced by stochastically modeled autonomous movements ($p = 0.5$, $\sigma = 0.2$ mm) according to (2.19). The covariance due to autonomous movements increases in time. Right: variance of autonomous movements for different values of the power p. If p is equal to 0.5, the variance increases linear in time. If p is less than 0.5, the increase in variance reduces with increasing time (settlement behavior)

where $\underline{z}_i^t$ represents the model imperfection of point i at time t; $E\{\underline{z}_i^t\} = 0$. The distance between the points i and j is given by l_{ij}. In (2.20), model imperfections are modeled with a power model in time and an exponential covariance function in space. The exponential covariance function in space models the deviations that have a spatial correlation of length L. Due to the depth of the gas fields of ~3 km, a correlation length of at least 3 kilometers is expected for the deformation signal. The correlation length was set to 4 kilometers in the estimation of subsidence due to gas extraction in 2003 (Schoustra, 2004), to cover all spatially correlated deformation signal. The power model in time takes random walk deviations into account that are caused by an under- or overestimation of the subsidence prognosis in time.

The functional and stochastic modeling contributes to the precision and reliability of a measurement technique for monitoring the deformation signal of interest. Therefore, model imperfections will be further addressed in Sect. 6.5 as a part of the quality assessment.

2.3.4 Deformation Analysis of Subsidence Due to Gas Extraction

Both the point-wise multi-epoch (Sect. 2.3.2) and the continuous spatio-temporal deformation analysis (Sect. 2.3.3) have been applied in the Netherlands by gas and oil companies for monitoring subsidence due to hydrocarbon production.

Point-wise multi-epoch deformation analysis has the advantage that it provides direct insight in the movements of individual benchmarks. An example of point-wise multi-epoch deformation analysis in Groningen is the analysis of subsequent benchmark heights obtained from leveling campaigns (de Heus et al., 1994). A disadvantage of this method is that the benchmark heights are dependent on the choice of the reference benchmark(s). Furthermore, 2% of the benchmarks disappear yearly which results in incomplete time series (Schoustra, 2004). Moreover, the spatio-temporal correlation of the deformation signal of interest is not utilized.

When modeling subsidence as a continuous spatio-temporal phenomenon (see Sect. 2.3.3), incomplete time series can easily be incorporated. Furthermore, due to the introduction of the spatio-temporal correlation of subsidence, outliers, identification errors and autonomous benchmark movements can be detected and removed in an automatic way. Finally, there is no dependency on the choice of the reference benchmark(s) due to the usage of the height difference measurements as the basic observations.

Methodologies that apply the continuous spatio-temporal deformation analysis concept are described by Kenselaar and Quadvlieg (2001) and Houtenbos (2004). The Subsidence Modeling (SuMo) concept of Kenselaar and Quadvlieg (2001) models the subsidence signal z as a (superposition of) ellipsoidal bowl(s) with decreasing linear benchmark velocities from the center of the bowl. The Subsidence Residual modeling (SuRe) concept of Houtenbos (2004) uses subsidence prognosis grids based on geomechanical modeling of the subsurface.

The mathematical framework underlying both SuMo and SuRe can be summarized as:

$$\underline{h}_{ij}^{t} = H_j^{t_0} - H_i^{t_0} + z_j^{t_0 t} - z_i^{t_0 t} + \underline{\delta h}_{ij}^{t} + \underline{\delta s}_j^{t} - \underline{\delta s}_i^{t} + \underline{\delta z}_j^{t_0 t} - \underline{\delta z}_j^{t_0 t}, \tag{2.21}$$

where $\underline{h}_{ij}^{t}$ is the spatial height difference observation between points i and j at time t, and $H_j^{t_0}$ and $H_i^{t_0}$ are the unknown initial heights. The functional model of subsidence due to gas extraction is denoted by z. When using prognosis grids, z is subtracted from the height difference observations, which results in subsidence residuals (SuRe). The stochastically modeled components are $\underline{\delta h}$ (measurement noise), $\underline{\delta s}$ (autonomous movements) and $\underline{\delta z}$ (subsidence model imperfections). In the SuRe methodology, the parameters of these stochastically modeled components are estimated through variance component estimation. Examples of stochastic parameters are variance factors, spatial correlation length and temporal power, see (2.20). An application of the SuRe concept including VCE is described in Sect. 6.5.4.2. Here, deformation components in Rotterdam are separated into autonomous movements and spatio-temporally correlated deformation signal.

Although continuous spatio-temporal deformation analysis has clear advantages, it is essential that the covariance function(s) that describe the model imperfections are adequate for the estimation of the signal of interest. If not, a risk exists of the attribution of displacement components to the wrong deformation cause. Hence, the link to geomechanics should be prominent in continuous spatio-temporal deformation analysis.

2.4 Conclusions

This chapter has summarized the geological and geomechanical properties of hydrocarbon reservoirs, and has focused specifically on the Groningen gas field in the Netherlands. Based on the reservoir properties and production scenarios, the subsidence pattern at ground level can be predicted. The resulting subsidence is dependent on the geometrical shape of the reservoir, the compaction coefficient, the thickness of the reservoir, the pressure drop in the reservoir, and the geomechanical behavior of the overburden. Several methodologies are available to predict subsidence: from analytical expressions as a function of a few reservoir parameters to finite element analyses that take the spatially varying geomechanical parameters in the subsurface into account.

Geodetic techniques, such as leveling, are applied to measure deformation at ground level. Therefore, the subsidence signal is discretizised by a set of measurement points, that are monitored at subsequent epochs. Two types of deformation analysis have been addressed: point-wise multi-epoch and continuous spatio-temporal deformation analysis. Point-wise multi-epoch deformation analysis estimates deformation in time for each benchmark individually. Continuous spatio-temporal deformation analysis incorporates the spatio-temporal correlation of the signal of interest. When the functional model of the deformation signal is not well known, model imperfections can be modeled stochastically. To conclude, the applied deformation analysis methodologies in the Netherlands have been summarized.

Chapter 3 will focus on Persistent Scatterer InSAR as a measurement technique that can be applied for deformation monitoring from a geodetic point of view. The overview of deformation estimation concepts that have been presented in this chapter will be applied in the interpretation of InSAR displacement estimates in Sects. 4.5 and 6.5, and in the integration of InSAR and leveling deformation estimates in Chap. 7.

Chapter 3
Persistent Scatterer InSAR

In Chap. 2, the physical mechanism of subsidence due to hydrocarbon production and deformation estimation methodologies have been discussed. This chapter introduces InSAR as a measurement technique for deformation monitoring.

InSAR utilizes phase difference observations between two radar acquisitions for the estimation of surface deformation. Various SAR missions are currently operational, such as ERS-2 and Envisat, acquiring SAR images that cover an area of 100×100 km with a 35 day time interval, at 800 km height, using C-band wavelength of 56 mm. Interferometric phase observations can be obtained with millimeter precision. However, besides the deformation signal of interest, the interferometric phase also contains contributions due to atmospheric signal delay, topography and orbital errors. Moreover, only the fractional phase is observed, which implies that the number of integer cycles from satellite to surface is unknown. Conventional InSAR can only be successfully applied in areas that do not suffer from significant surface changes in time, whereas the deformation signal has such magnitude that error sources are negligible.

Subsidence due to hydrocarbon production in the Netherlands has a small magnitude (<1 cm/year) and a large spatial extent (the extent of the Groningen gas field is $\sim$30 km). Furthermore, the area of interest has an agricultural character and is subject to surface changes (temporal decorrelation). Hence, an accurate estimation of the error sources, such as atmospheric disturbances, is very important. Therefore, a time series of SAR acquisitions is utilized to detect reliable measurement targets, based on their reflectivity behavior through time: the so-called Persistent Scatterers (PS) (Ferretti et al., 2000).

In this chapter, the estimation of deformation from PS phase observations is explained (functional model, stochastic model, and estimation procedure). It starts with an overview of interferometric processing. Subsequently, the selection of Persistent Scatterer candidates is addressed in Sect. 3.2. Several strategies have been developed to estimate deformation from the PS observations (Ferretti et al., 2001; Berardino et al., 2002; Kampes, 2005). In this research, the Delft implementation of Persistent Scatterer InSAR (DePSI) has been applied. The mathematical framework underlying DePSI is explained in Sect. 3.4. The functional and stochastic models for PSI are considered from a theoretical point of view in this chapter. The actual application of DePSI for subsidence monitoring in the Netherlands will be addressed in Chap. 6.

3.1 Interferometric Processing

SAR images consist of a grid of complex observations that can be addressed to single resolution cells. These complex observations stem from the superposition of all

V.B.H. (Gini) Ketelaar, *Satellite Radar Interferometry*,
Remote Sensing and Digital Image Processing 14,

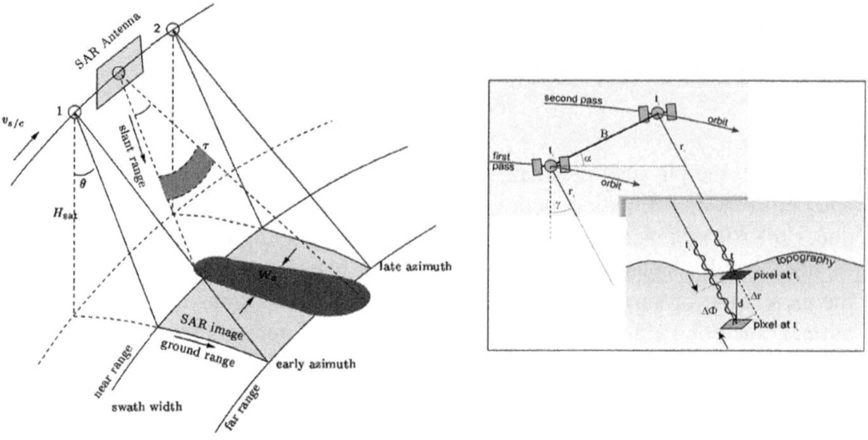

Fig. 3.1 Left: image geometry of a SAR acquisition (Hanssen, 2001). The satellite velocity v_s is approximately 7 km/s. The dark grey area indicates the footprint of a single pulse. The total coverage of a SAR scene, between early and late azimuth, and near and far range, is depicted in light grey. Right: repeat-pass InSAR and detection of deformation by means of phase difference observations

scattering targets within the resolution cell. The range resolution of a SAR acquisition depends on the system's bandwidth (Hanssen, 2001). ERS-2 for example has a range bandwidth of 15.55 Mhz, which corresponds with a range resolution of 9.6 m. In azimuth direction, the Doppler bandwidth is utilized for optimizing the azimuth resolution. The Doppler bandwidth of ERS-2 is ~1378 Hz. The Pulse Repetition Frequency (PRF) is a higher, 1680 Hz, to fully sample the Doppler spectrum.

The pixel spacing of a SAR image is closely related to the resolution. The pixel spacing is determined by the sampling rate. For ERS-2, the range sampling rate is 18.96 Mhz. The ERS-2 pixel spacing in range direction is 7.9 m, which corresponds with ~20 m in ground range. The ERS-2 azimuth spacing is 4 m. In this book, the term pixel is used as the area covered by the pixel spacing, although strictly speaking a pixel is a infinitesimally small point that represents the resolution cell (Hanssen, 2001).

Each object on earth is observed by consecutive pulses, appearing in different range bins in the raw data. By means of focusing, these observations are shifted to a common range bin. In this research, the input in the interferometric processing chain are the focused Single-Look Complex (SLC) images.

Each complex observation can be converted into an *amplitude* and a *phase* observation. The amplitude represents the strength of the reflection towards the sensor from the resolution cell. An *interferogram* is computed by the complex multiplication of the observations in each resolution cell from two coregistered SAR images: a 'master' and a 'slave' image. The interferometric phase differences are the actual observations from which earth surface deformation is estimated.

The computation of interferograms in this research has been performed by the Delft Object-oriented Radar Interferometric Software (Doris) (Kampes and Usai, 1999). The key steps are summarized in Fig. 3.2.

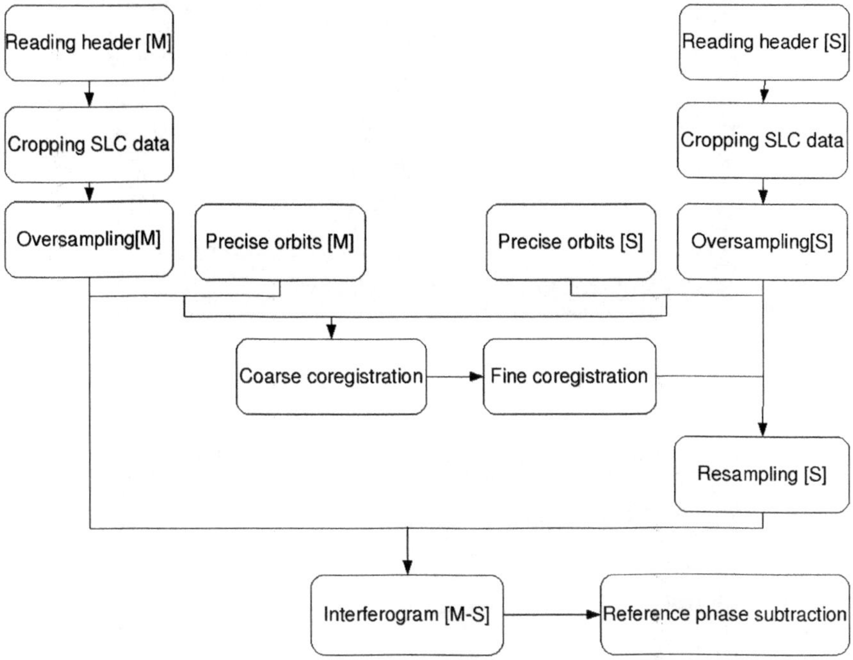

Fig. 3.2 Schematic overview of interferometric processing. A master (M) and a slave acquisition (S) are oversampled and coregistered. The interferogram is computed from the complex multiplication of the master image and the resampled slave image. In the last stage, the reference phase of the ellipsoid is subtracted

3.1.1 Oversampling

After reading the SLC data and the precise orbits, the SAR images are oversampled with a factor 2 prior to the coregistration and the formation of interferograms. Oversampling with a factor 2 is performed to avoid aliasing in the complex multiplication of the SAR images. Since the equivalent of a multiplication in the space domain is a convolution in the frequency domain, the spectrum length (both in azimuth and range) will be doubled after complex multiplication of two SAR images. The effect of oversampling is illustrated in Fig. 3.3 and Fig. 3.4. For ERS, the pulse repetition frequency is 1680 Hz, while the azimuth bandwidth is 1378 Hz. After oversampling, the pulse repetition frequency is 3360 Hz, which avoids aliasing in the complex multiplication. In the determination of the minimal oversampling factor, the Doppler centroid frequency shift has to be compensated. The Doppler centroid frequency is the center frequency of the passage of a scatterer through the antenna beam. ESA's SLC images are zero-Doppler processed, which means that the corresponding satellite position can be found by an orthogonal projection on the precise orbits. However, during the actual acquisition, the true Doppler centroid is never aimed exactly perpendicular to the flight direction. Therefore, the SAR data spectrum is shifted in azimuth direction. An overview of this mechanism can be found

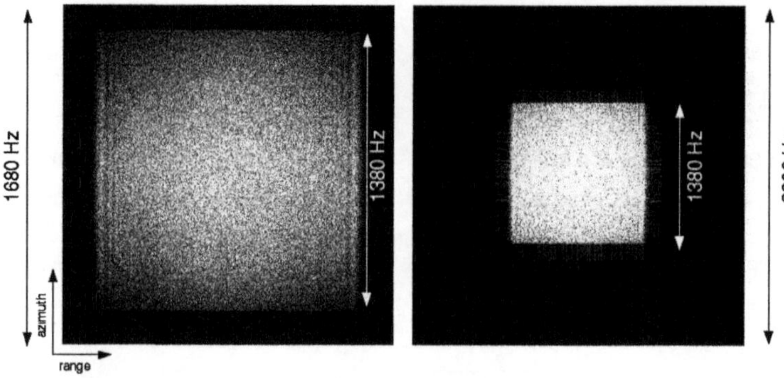

Fig. 3.3 Original SAR image spectrum (left) and after oversampling with a factor of 2 (right)

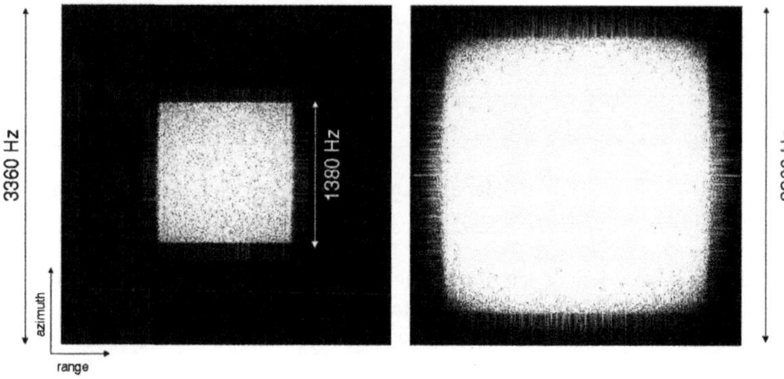

Fig. 3.4 SAR image spectrum after oversampling with a factor of 2 (left) and after complex multiplication (right). The size of the spectrum is doubled after oversampling; due to the oversampling, aliasing does not occur

in Hanssen (2001). For ERS-2 acquisitions with high Doppler centroid frequencies (>2000 Hz), oversampling with a factor 2 may not be sufficient to avoid aliasing. An oversampling factor of 2 avoids aliasing of Doppler shifts up to 990 Hz; a factor 4 covers Doppler shifts up to 2670 Hz. Some ERS-2 acquisitions exhibit even higher Doppler deviations, requiring even higher oversampling factors.

The radar coordinates of an oversampled image are at sub-pixel level with respect to the original sampling rate. This implies that a target's location in the radar coordinate system can be more precisely determined in an oversampled image.

3.1.2 Coregistration

The oversampled SLC images together with the precise orbits (Scharroo and Visser, 1998; Doornbos and Scharroo, 2004) form the input for the coregistration of the master and the slave image. Coregistration is a crucial step. If its precision is not

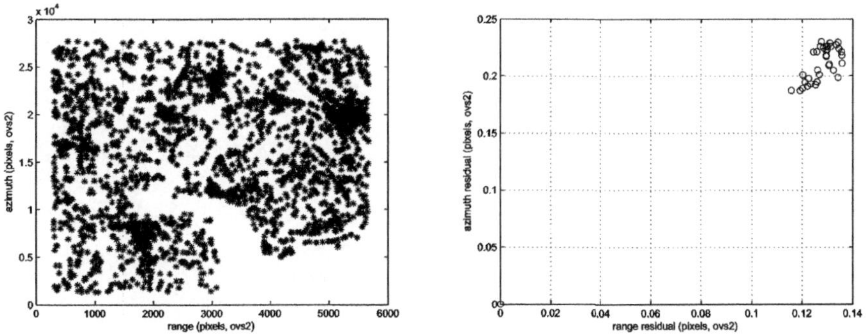

Fig. 3.5 Spatial distribution of observations for the estimation of the coregistration polynomial (left) and standard deviation of the coregistration residuals per acquisition (right). The coregistration windows are evenly distributed over the area, also in rural parts. The part of the image that does not contain window locations is covered with water. The standard deviation of the coregistration residuals is ~0.1 and ~0.2 oversampled pixels in range and azimuth direction respectively. This is equivalent to ~0.05 and ~0.1 pixels in range and azimuth direction in the original resolution

sufficient, it deteriorates the PS selection process and decreases the PS phase observation precision. In the northeastern part of the Netherlands, the height differences are small (<30 meters). Here, coregistration can be performed by a second degree polynomial. For the estimation of an accurate coregistration polynomial, it is essential that the coregistration windows are evenly distributed over the area of interest. Especially in rural areas that are affected by temporal decorrelation, a proper choice of coregistration windows is required. Hence, a large number of candidate windows have been placed around evenly distributed local amplitude maxima, which are assumed to correspond with terrain features that are stable in time.

The coregistration polynomial is estimated from a set of corresponding locations in master and slave acquisitions that are acquired from correlation optimization between the (oversampled) candidate windows. To remove outliers, a geodetic testing procedure is applied (Teunissen, 2000b; Kampes, 2005). In this step, it is important that the size of the test, which determines the likelihood of rejecting a good observation (type I error), is not set too strict. If too many observations are incorrectly rejected, the spatial coverage of the observations decreases, which results in inaccurately extrapolated coregistration vectors in rural areas with few PS candidates.

Figure 3.5 shows an example of the spatial distribution of the accepted observations in the estimation of the coregistration polynomial. The locations with a high observation density correspond with the city areas. The areas that are covered with water do not contain any observations, due to temporal decorrelation. The standard deviation of the coregistration residuals is approximately 0.1 and 0.2 pixel in range and azimuth direction respectively (oversampling factor 2). By depicting the spatial distribution of the coregistration observations and their residuals, images with inaccurate coregistration are easily detected.

Fig. 3.6 Interferogram before (left) and after (right) subtraction of the reference phase. The temporal and perpendicular baseline are 140 days and 166 m respectively. The repetitive line pattern (left) depicts the phase contribution due to the ellipsoidal shape of the earth. After subtraction of the reference phase, only urban areas appear coherent in the interferogram (right). Large parts of the image are noisy due to temporal decorrelation. The SAR data has been provided by ESA

3.1.3 Interferogram Computation

After the estimation of the coregistration polynomial, the slave image is resampled to the master image geometry. Subsequently, the interferogram is computed by complex multiplication of the master image and the resampled slave image observations. In the last step, the interferometric phase contribution due to the ellipsoidal shape of the earth is subtracted. Figure 3.6 depicts the interferogram before and after reference phase subtraction. The repetitive fringe pattern in the interferogram before subtraction of the reference phase depicts the phase contribution due to the ellipsoidal shape of the earth. After subtraction of the reference phase, the urban areas that contain coherent man-made features and the rural areas that suffer from temporal decorrelation can be distinguished. The interferometric phase not only represents surface deformation, but contains contributions due to atmospheric disturbance, topographic height differences, and residual orbital errors as well. To estimate deformation in the presence of other phase contributions, a network of measurement points (the Persistent Scatterers) will be selected by using a time series of SAR acquisitions. The selection of these measurement points is addressed in the next section.

3.2 Persistent Scatterer Selection

Not all phase observations in an interferogram contain useful information. The earth surface covered by a resolution cell may change in time, targets may not reflect back in the satellite direction, or they may not have the physical properties to be observed with a high enough Signal-to-Noise Ratio (SNR). In order to select coherent, i.e. interpretable, measurement points, candidate targets are selected that have a strong and consistent reflection in time (the Persistent Scatterers (Ferretti et al., 2000)).

These targets can often be attributed to a single physical object ('man-made feature') that acts as a dominant scatterer within the resolution cell. Many of them behave like point scatterers, and can be observed from a wide range of viewing angles. These point scatterer observations are therefore less sensitive for the effective (perpendicular) baseline between the master and slave acquisition. A Persistent Scatterer does not necessarily have to be a dominant point scatterer within the resolution cell. Distributed scatterers, such as rocks, can act as PS as well, although with a different dependency on the viewing geometry, i.e. the perpendicular baselines.

This section addresses the selection of PS candidates. It explains the methodologies to select targets with a high likelihood of coherent phase behavior from a time series of SAR acquisitions.

3.2.1 Identification Methods of PS Candidates

In order to estimate all interferometric phase contributions (surface deformation, topographic heights, atmospheric disturbances, etc.) and maintain an efficient algorithmic approach, a first order network is constructed from the potentially most reliable PS candidates. Since the interferometric phase observations are wrapped and the phase contribution due to the unknown deformation is contaminated by several 'error sources', the amplitude observations of the PS candidates are utilized. Existing methodologies can be subdivided into three groups that use the following parameterization to select PS candidates:

1. Signal-to-Clutter Ratio (SCR) (SCR, 1993),
2. normalized amplitude dispersion (D_a) (Ferretti et al., 2001), and
3. supervised classification (Humme, 2007).

SCR estimation is based on the assumption that a PS observation consists of a deterministic signal that is disturbed by random circular Gaussian distributed clutter, see Fig. 3.7. The deterministic signal stems from the dominant scatterer within the resolution cell. The clutter reflects the distributed scatterers in the surroundings of the point scatterer. The relation between SCR and the phase variance is defined as (SCR, 1993):

$$\mathrm{SCR} = \frac{s^2}{c^2}; \qquad \sigma_{\psi}^2 = \frac{1}{2 \cdot \mathrm{SCR}} (\mathrm{rad}), \tag{3.1}$$

where σ_{ψ}^2 is the phase variance of a single SAR observation, s represents the amplitude of the dominant scatterer and c the clutter in the surroundings.

A scatterer with a high SCR through time is labeled as a PS candidate. SCR assumes stationary stochastic behavior of the surroundings of the scatterer, which may not be valid, especially in urban areas that contain multiple scatterers at a short distance that are likely to interfere. Practical implementation of SCR estimation is therefore not straightforward. Automatic distinction of two nearby PS candidates requires high flexibility of SCR estimation windows and signal edge detectors.

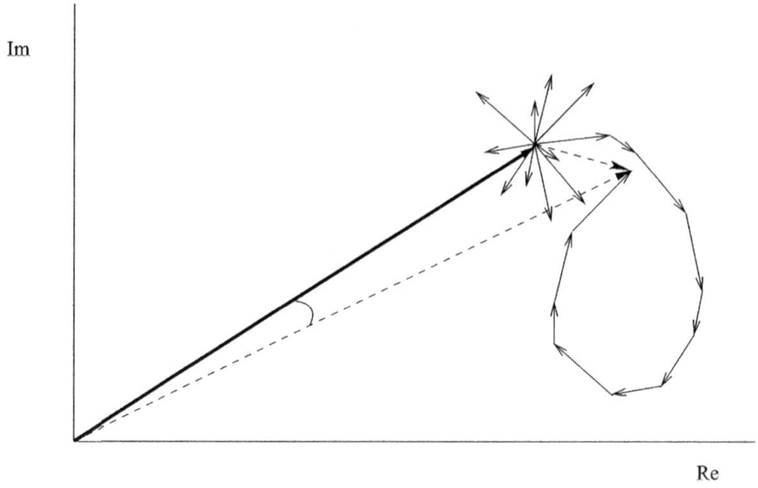

Fig. 3.7 Complex PS observation that consists of a deterministic signal with superposed Gaussian distributed clutter (solid arrow lines). The deterministic signal represents the dominant scatterer within the resolution cell. The small dashed arrow line indicates the resultant of the clutter signal, which imposes the uncertainty on the phase observation of the dominant scatterer

The normalized amplitude dispersion method (Ferretti et al., 2001) performs an amplitude time series analysis, in stead of a spatial analysis. Each pixel is quantified by the ratio between the dispersion of the amplitude σ_a and the mean μ_a of the amplitudes through time:

$$D_a = \frac{\sigma_a}{\mu_a}. \tag{3.2}$$

Point scatterers with a low normalized amplitude dispersion have a low phase dispersion. Hence, they are selected as PS candidates. A direct relation exists between the normalized amplitude dispersion D_a and SCR (SCR, 1993):

$$D_a = \frac{1}{\sqrt{2 \cdot \mathrm{SCR}}}. \tag{3.3}$$

A typical threshold for the normalized amplitude dispersion is 0.25 (Ferretti et al., 2001), which corresponds with a SCR of 8.

PS candidate selection based on SCR and normalized amplitude dispersion are using the amplitude observations. Phase stability as a selection criterium for PS candidates has been investigated as well (Hooper et al., 2004). The phase stability is analyzed under the assumption that deformation is spatially correlated. The phase observations of neighboring PS candidates are averaged, and those with the lowest residual noise are selected. However, with the typical land use in the Netherlands (scattered cities and villages separated by agricultural and vegetated areas) only a small percentage of the pixels in an interferogram contains Persistent Scatterers. Hence, also for this method, a pre-selection is made based on the amplitude observations.

Recently, supervised classification to select PS candidates has been investigated as well (Humme, 2007). In areas where the distribution of PS is crucial for the estimation of the signal of interest, the possibility of omitting a PS candidate is reduced by manual selection of pixels that contain the reflection of man-made features.

For the selection of PS candidates, it is essential that the amplitude observations are not affected by satellite system characteristics and viewing geometry. To obtain the amplitude observations that represent solely the physical characteristics of the PS candidates, SAR amplitude calibration is performed. The next section introduces a new method that does not perform the amplitude calibration explicitly, and hence saves a considerable amount of computation time.

3.2.2 *Pseudo-calibration*

For an unbiased selection of PS candidates based on amplitude, SAR calibration is performed to isolate the amplitude observations corresponding with physical PS properties from amplitude variations due to viewing geometry and sensor characteristics. A method to perform SAR calibration is the ESA ERS SAR calibration (Laur et al., 2002). This SAR calibration method quantifies the satellite system characteristics. The amplitude observations are corrected for the varying system characteristics by means of a multiplication factor. This multiplication factor is built up from constant factors per acquisition (calibration constant, antenna pattern gain, replica pulse power); factors that depend on viewing geometry (range, incidence angle) and a factor that varies over the entire image (power loss).

The ESA SAR calibration method is useful for thematic interpretation purposes, where amplitude values have to be compared over the entire SAR image. The variation of the location of a Persistent Scatterer within a time series of SAR images is nevertheless limited. For PS selection, there is no need to be able to compare amplitude values between near and far range. It is sufficient to determine calibration factors that are valid in the near surroundings of the PS. Moreover, uncalibrated amplitude observations of potential PS candidates themselves can be utilized for the estimation of the calibration factors, since they correspond with man-made features that are not affected by temporal decorrelation. This has resulted in the development of empirical calibration methods, that are sensor independent (Bovenga et al., 2002; Cassee, 2004).

The newly developed calibration method that is described in this section goes one step further: it does not perform empirical calibration explicitly but evaluates if the PS candidate would have been selected if calibration would have been performed. It is referred to as pseudo-calibration. This method has the advantage that it saves computation time and storage space for calibrated images, especially since the amount of PS is generally a small percentage of the full image (1–2% or lower) and only their interferometric phase observations are used for the estimation of the deformation signal. In this section, the mathematical framework for pseudo-calibration is described and its application to real data.

3.2.2.1 Mathematical Model

Empirical SAR calibration is implemented as a Gauss-Markov model that consists of a functional and stochastic model. The observations are the amplitudes of potential PS candidates; the unknowns are the amplitude multiplication factors per image (the calibration factors).

Since the sensor characteristics and viewing geometry vary smoothly over a SAR image, the image is subdivided into patches where the amplitude multiplication factors are assumed to be constant. In each of these patches, potential PS candidates are selected based on the normalized amplitude dispersion that is computed from the stack of uncalibrated images. For a selection of P neighboring potential PS candidates, a multiplication factor c_k is estimated with respect to a reference image for all K SAR acquisitions. The functional and stochastic model read:

$$E\{\underline{y}\} = E\{\underline{a}_p^k\} = c_k a_p^{\text{ref}}; \qquad D\{\underline{y}\} = \sum_{p=1}^{P} \sigma_{a_p}^2 Q_p, \tag{3.4}$$

where $\underline{a}_p^k$ is the amplitude observation for PS candidate p in image k with its variance $\sigma_{a_p}^2$. The unknown amplitude for PS candidate p free from variations due to sensor characteristics and viewing geometry is represented by a_p^{ref}. As the calibration factors can only be estimated relatively, one calibration factor is fixed at the value 1. Solving the system of equations from (3.4) results in a set of calibration factors that are utilized for pseudo-calibration.

The stochastic model from (3.4) is not very well known a-priori. It is assumed that the amplitude variance reflects the physical properties of the PS candidates superposed on random noise due to viewing geometry (incidence and squint angle). Its initial value is derived from the amplitude dispersion (amplitude variance) in the uncalibrated stack.

Based on the adjustment residuals, variance factor(s) to update the stochastic model are estimated by variance component estimation (VCE) (Teunissen, 1988), see Sect. 2.3.1. In the SAR calibration validation either one variance factor $\hat{\underline{\sigma}}^2$ for the entire variance matrix can be estimated, or a variance factor per PS candidate: $\hat{\underline{\sigma}}_1^2 \ldots \hat{\underline{\sigma}}_P^2$. Due to the redundancy that is required for a precise estimation of variance factors, it is chosen to update the variance-covariance matrix with one variance factor. The relative weights between the PS candidates stem from the amplitude dispersion in the uncalibrated stack.

3.2.2.2 Testing Observational Errors

Erroneous observations, such as amplitudes of non-stable potential PS candidates, influence the estimation of the calibration factors. Hence, tests are performed to trace these erroneous measurement points and remove them from the dataset. A null hypothesis H_0 is compared to an alternative hypothesis H_A (Teunissen, 2000b):

$$H_0 : E\{\underline{y}\} = c_k a_p^{\text{ref}} \quad \text{versus} \quad H_A : E\{\underline{y}\} = c_k a_p^{\text{ref}} + C_y \nabla. \tag{3.5}$$

The *point test* is applied to trace false potential PS candidates in the estimation of the calibration factors. The point test is an integrated test on all amplitude observations of a single PS candidate. Its C matrix reads:

$$C_p = \left[0 \ldots 0 \, I_p \, 0 \ldots 0\right]^T, \tag{3.6}$$

where I_p assumes errors on all amplitude observations of PS p.

3.2.2.3 Threshold Tuning

By application of the calibration factors, a time series of calibrated images can be obtained. Subsequently, one of the identification methods from Sect. 3.2.1 can be chosen to perform the selection of PS candidates. This requires the computation and storage of adapted amplitude observations. Pseudo-calibration omits this step: it tests if the PS candidates would have been selected if the images would have been calibrated.

Pseudo-calibration adapts the normalized amplitude dispersion threshold for the ideal situation (no amplitude variations due to sensor characteristics) to the equivalent threshold for a stack of non-calibrated images. We refer to this as *threshold tuning*. It is based on a Monte-Carlo simulation of random samples with the normalized amplitude dispersion threshold. These samples are multiplied with the calibration factors and a new normalized amplitude dispersion threshold is estimated for the non-calibrated situation. Figure 3.8 depicts the modified threshold as a function of normalized amplitude dispersion for a stack of SAR images.

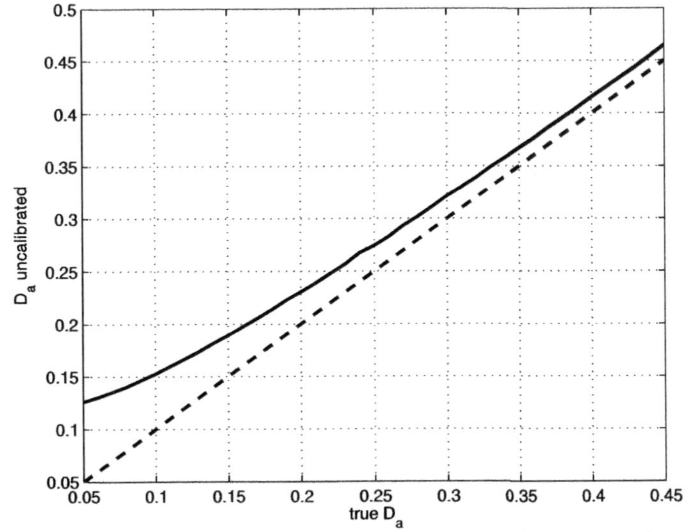

Fig. 3.8 Threshold tuning for pseudo-calibration: based on the estimated calibration factors, a corresponding threshold is computed for the non-calibrated stack (solid). The dashed line represents the true normalized amplitude dispersion

Since the PS amplitudes are only used for PS selection, this is a simplified but fast method for PS candidate detection. Subsequently, based on the phase difference observations, it is determined if a PS candidate is accepted or rejected.

3.2.2.4 Application of Pseudo-calibration

Performance analysis of pseudo-calibration and comparison to other methods has been done based on the phase history of the selected potential PS. The phase residuals per arc between two potential PS have been parameterized in a phase coherence measure (Ferretti et al., 2001):

$$\gamma = \left| \frac{\sum_{i=1}^{N} e^{j\Delta w_i}}{N} \right|, \tag{3.7}$$

where N is the number of interferograms and Δw_i are the phase residuals between two PS. A PS candidate is accepted if it is part of at least two arcs that exceed the phase coherence threshold.

For a test area of 8×6 kilometers, the ESA ERS SAR calibration method has been compared to pseudo-calibration. A stack of 73 ERS-1 and ERS-2 images were analyzed in the selection of a sparse grid of PS candidates for the estimation of (residual) topography and deformation. For each grid cell of 200 × 200 meters, the best potential PS with the lowest normalized amplitude dispersion below the threshold of 0.25 were selected. Figure 3.9 shows the detected potential PS for both ESA ERS SAR calibration and pseudo-calibration. Table 3.1 lists the number of detected potential PS, the percentage of rejected potential PS based on phase behavior (coherence threshold 0.75), and the accepted potential PS in common.

These results show that for PSI, pseudo-calibration is an alternative for the calibration of full images based on physical sensor parameters. Pseudo-calibration detects more PS candidates, while the false detection rate based on the phase observations is lower.

3.3 Persistent Scatterer Phase Observations

Now the PS candidates have been selected, their interferometric phase observations are utilized for the confirmation of the current PS selection and the estimation of the deformation signal of interest. This section focuses on the interferometric combinations and the network design. The functional model will be addressed in Sect. 3.4.

3.3.1 Master Selection

An interferogram results from complex multiplication of a master and a slave image. For K SAR acquisitions, $(K-1)$ independent interferometric combinations

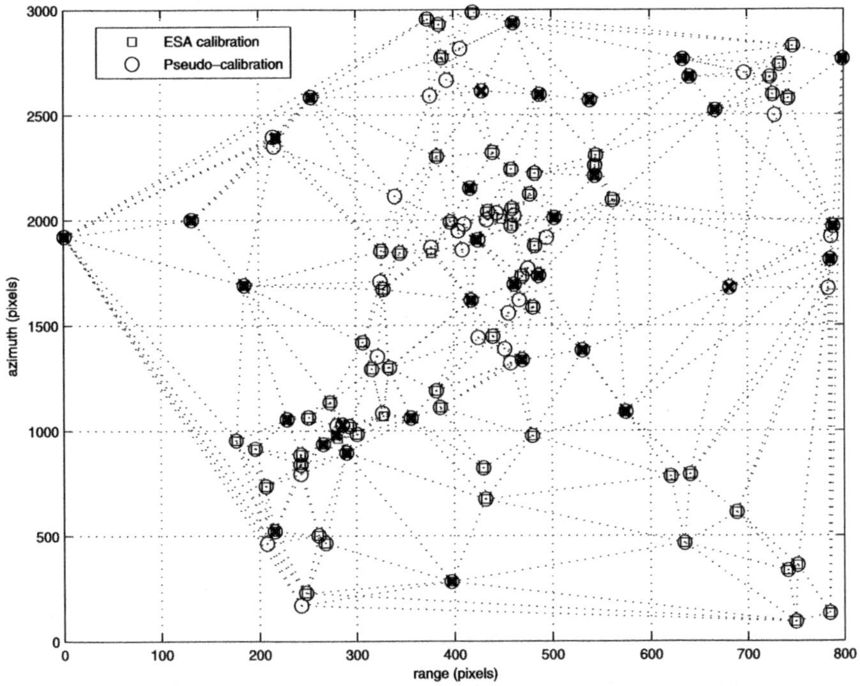

Fig. 3.9 Detected PS candidates for ESA ERS SAR calibration (squares) and pseudo-calibration (circles). Rejected points based on phase coherence are marked with a black cross. Pseudo-calibration, that does not perform amplitude calibration explicitly, can replace calibration based on sensor characteristics

Table 3.1 Comparison of ESA ERS SAR calibration and pseudo calibration

	# PS	Rejected	Common PS
ESA calibration	58	35%	81%
Empirical validation	93	21%	81%

between two images can be formed. Common practice is to create a stack of interferograms that have a common master. However, in fact the interferometric combinations can be chosen in an arbitrary way (for example by minimizing the perpendicular baselines between the combinations). For the same set of PS, the deformation estimates should be invariant of the choice of the interferometric combinations.

For a single master stack, the master image is selected based on the stack coherence (Kampes, 2005). The stack coherence is a function of perpendicular baseline $B_{\perp}$, temporal baseline T and Doppler centroid frequency f_{dc}:

$$\gamma^m = \frac{1}{K}\sum_{k=1}^{K} g(B_{\perp}^{k,m}, B_{\perp_{\max}}) \cdot g(T^{k,m}, T_{\max}) \cdot g(f_{\mathrm{dc}}^{k,m}, f_{\mathrm{dc}_{\max}}), \tag{3.8}$$

where

$$g(x,c) = \begin{cases} 1 - |x|/c & \text{for } |x| \leq c, \\ 0 & \text{for } |x| > c, \end{cases} \tag{3.9}$$

and m refers to the master acquisition and k to the slave acquisitions. Plausible values for $B_{\perp_{\max}}$, $T_{\max}$ and $f_{\mathrm{dc}_{\max}}$ are respectively 1500 meters, 15 years and 1380 Hz (the azimuth bandwidth). As point scatterer reflections are consistent over a wide range of viewing angles, the maximum perpendicular baseline can be chosen larger than the critical baseline $B_{\perp,\mathrm{crit}}$, which is the baseline causing a spectral shift equal to the range bandwidth (for ERS 1.1 km). PS refer to features that exhibit minimal temporal decorrelation. Hence, the temporal window can be set in order to cover the temporal range of the entire stack.

3.3.2 Double-difference Observations

A PSI measurement $\underline{\varphi}_{p}^{ms}$ is an interferometric phase difference in time between master m and slave s, for PS p. Only the fractional phase is measured and not the integer number of cycles from satellite to the earth surface: the phase observations are 'wrapped'. The first 'interpretable' PSI observation is the double-difference $\underline{\varphi}_{pq}^{ms}$, between master and slave and between PS p and q (Hanssen, 2004). The double-difference is both a temporal and a spatial difference. This implies that PSI observations require a spatial and a temporal reference (basis): one acquisition time and one PS. For the same set of PS, the deformation estimates should be invariant of the choice of the spatial and temporal reference. The number of independent double-differences that can be formed from the original phase observations is equal to $(K-1)\cdot(P-1)$, for K SAR acquisitions and P PS. In this study, double-differences will be denoted as $\underline{\varphi}_{pq}^{k}$, where $k = 1\ldots(K-1)$, with $(K-1)$ the number of independent interferometric combinations.

The number of SAR acquisitions and PS determine the number of independent double-differences that can be formed. The interferometric (temporal) combinations have been discussed in Sect. 3.3.1. An option for the spatial distribution is the star network that is depicted in Fig. 3.10. In the star network, the spatial differences refer to one PS. Similar to the interferometric combinations, this is not a requirement. From P PS, $(P-1)$ arbitrary independent combinations can be formed. For example, the SBAS methodology (Berardino et al., 2002) could be utilized, optimizing the perpendicular baselines for the stack of acquisitions to limit spatial decorrelation. Taking into account that certain targets only act as a Persistent Scatterer in a limited time window, the PSI estimation can be optimized as a function of both perpendicular and temporal baseline.

A disadvantage of the star network is that erroneous PS candidates cannot be detected. The star network contains less and longer arcs compared to a network based on the Delaunay triangulation, see Fig. 3.10. The disadvantage of longer arcs is that the phase difference observations contain more atmospheric signal and hence are

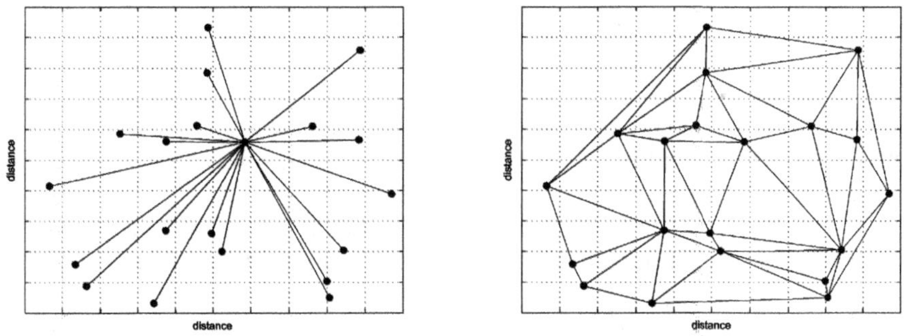

Fig. 3.10 Spatial PS network configurations: star network (left) and the network based on Delaunay triangulation (right). The spatial differences in the star network are linearly independent. The network based on Delaunay triangulation contains dependent spatial differences. Although the dependent combinations do not add information in the estimation of the deformation signal, they are used to detect erroneous PS candidates

more sensitive for unwrapping errors. Using the network based on the Delaunay triangulation, phase unwrapping and parameter estimation are performed per arc by the application of the functional model that will be described in Sect. 3.4. Subsequently, misclosure tests are performed. In this way, PS candidates can be rejected when the misclosures of the connecting arcs are not all 0. The redundant network is used to trace and remove erroneous PS candidates. For estimation of the deformation signal, only $(P-1)$ independent spatial combinations can be utilized.

3.4 PSI Estimation

This section describes the Delft PSI estimation methodology (DePSI), that is based on geodetic adjustment and testing techniques. Successively, the functional model, Integer Least-Squares estimation, and the stochastic model are addressed. The section concludes with the DePSI estimation sequence that starts with a *first order* network that is subsequently densified.

3.4.1 Functional Model

Interferometric double-difference observations contain contributions due to deformation, (residual) topography, atmospheric signal, orbit inaccuracies, the position of the dominant scatterer within the resolution cell, and measurement noise. The position of the dominant scatterer within the resolution cell is referred to as sub-pixel position. The observations are wrapped, meaning that only the fractional phase is observed and the number of integer cycles from satellite to earth surface is unknown. The wrapped interferometric phase φ relates to the unwrapped interferometric phase

$\underline{\varphi}^{\text{unw}}$ as:

$$\underline{\varphi}^{\text{unw}} = \underline{\varphi} + 2\pi \cdot a, \tag{3.10}$$

where $a \in Z$ is the integer ambiguity. In this book, both the wrapped and unwrapped phase observations will be denoted as $\underline{\varphi}$. In the context, it will clearly stated if $\underline{\varphi}$ refers to wrapped or unwrapped phase observations.

Taking all phase contributions into account including its wrapped nature, the PSI system of observation equations reads, see e.g., Kampes (2005):

$$\begin{aligned}\underline{\varphi}_{ij}^{k} = -2\pi a_{ij}^{k} - \frac{4\pi}{\lambda}\frac{B_i^{\perp}}{R_i^m \sin\theta_i^m} H_{ij} - \frac{4\pi}{\lambda} D_{ij} + \frac{4\pi}{\lambda}\frac{B_i^{\perp}}{R_i^m \tan\theta_i^m}\eta_{ij}^{m} \\ + \frac{2\pi}{v}(f_{\text{dc},i}^{m} - f_{\text{dc},i}^{s})\xi_{ij}^{m} + \mathrm{f}_{\varphi_{\text{orbit}}}(\xi_{ij}^{m}, \eta_{ij}^{m}) + \underline{\varphi}_{ij\,\text{defo}}^{k} + \underline{\varphi}_{ij\,\text{atmo}}^{k} + \underline{n}_{ij}^{k},\end{aligned} \tag{3.11}$$

where:

$\underline{\varphi}_{ij}^{k}$	double-difference phase observation,
$B_i^{\perp}, R_i^m, \theta_i^m$	perpendicular baseline, range and incidence angle for PS i,
a_{ij}^{k}	integer ambiguity between PS i and PS j,
H_{ij}	(residual) topographic height between PS i and PS j,
D_{ij}	deformation between PS i and PS j,
ξ_{ij}^{m}	sub-pixel position in azimuth direction,
η_{ij}^{m}	slant-range sub-pixel position,
η, ξ	range and azimuth radar coordinates,
v	satellite velocity,
$f_{\text{dc},i}^{m}, f_{\text{dc},i}^{s}$	Doppler centroid frequency of master and slave acquisition,
$\mathrm{f}_{\varphi_{\text{orbit}}}(\xi_{ij}^{m}, \eta_{ij}^{m})$	(residual) orbital trend as a function of radar coordinates,
$\underline{\varphi}_{ij\,\text{defo}}^{k}$	residual deformation signal,
$\underline{\varphi}_{ij\,\text{atmo}}^{k}$	(residual) atmospheric signal,
$\underline{n}_{ij}^{k}$	measurement noise.

The noise component n, the measurement noise, is partly determined by the physical PS properties. Furthermore, it will contain components due to processing induced errors (coregistration, interpolation).

Steady-state subsidence due to gas extraction at a constant production rate can usually be modeled by a linear displacement rate: $T^k v_{ij}$, where T^k represents the temporal baseline and v_{ij} a constant velocity. Deviations from the linear model that have a bounded correlation length can be incorporated in the stochastic model that is discussed in Sect. 3.4.3.

The system of equations from (3.11) is underdetermined. Each double-difference observation has its own unknown ambiguity besides the other unknown parameters that have to be estimated (topographic height, deformation, etc.). A way to make the system of equations of full rank is by adding pseudo-observations. Consider a simplified system of equations in which the unknown parameters are reduced to the topographic heights H and the deformation parameters D. Adding pseudo-observations for the unknown parameters leads to the following system of equations (Hanssen, 2004):

$$E\begin{bmatrix}\underline{\varphi}_{ij}^{k}\\ \underline{d}\\ \underline{h}\end{bmatrix}=\begin{bmatrix}-\frac{4\pi}{\lambda} & -\frac{4\pi}{\lambda}\frac{B_i^{\perp}}{R_i^m\sin\theta_i^m} & -2\pi\\ 1 & 0 & 0\\ 0 & 1 & 0\end{bmatrix}\begin{bmatrix}D\\ H\\ a\end{bmatrix};\qquad D\begin{bmatrix}\underline{\varphi}_{ij}^{k}\\ \underline{d}\\ \underline{h}\end{bmatrix}=\begin{bmatrix}Q_{\varphi} & 0 & 0\\ 0 & \sigma_d^2 & 0\\ 0 & 0 & \sigma_h^2\end{bmatrix}. \tag{3.12}$$

The pseudo-observations are the initial values of the unknowns to be estimated. Their variance in the stochastic model is chosen in such way that the estimates can cover the entire physical range of values. Note that the variance-covariance matrix of the double-difference phase observations Q_{φ} is a full matrix, see Sect. 3.4.3.

Besides the lack of redundancy in (3.11), there is a clear linear dependency between the topographic height and the range sub-pixel position. Considering one arc between two PS i and j, it can be shown that the functional relation with the phase observation is dependent via multiplication of the height term with the cosine of the incidence angle:

$$-\frac{4\pi}{\lambda}\frac{B_i^{\perp}}{R_i^m\sin\theta_i^m}\cdot\cos\theta_i^m+\frac{4\pi}{\lambda}\frac{B_i^{\perp}}{R_i^m\tan\theta_i^m}=0. \tag{3.13}$$

This implies that the range sub-pixel position highly influences the (residual) height estimation. The range sub-pixel position can be estimated by Point Target Analysis (Werner et al., 2003) using the amplitude observations. Another option is to exclude the range sub-pixel position from the system of equations. The effect of the accuracy of the range sub-pixel position on the estimation of the deformation signal is further addressed in Sect. 4.2.1.

3.4.2 Integer Least-Squares Estimation

The system of equations from (3.11) contains integer parameters: the phase ambiguities. To each fractional phase observation, an integer number of phase cycles can be added or subtracted. Figure 3.11 shows the ambiguity in the displacement estimates, which is equal to half of the wavelength.

A system of equations that contains both integer and real-valued (float) unknowns can be solved by Integer Least-Squares estimation (ILS) (Teunissen, 2001b). The system of observation equations reads:

$$\underline{y}=Aa+Bb+\underline{e}, \tag{3.14}$$

where $\underline{y}$ is the vector of observations, a are the unknown integer parameters, and b the unknown float parameters. The residual vector $\underline{e}$ consists of model imperfections $\underline{s}$ and measurement noise $\underline{n}$. In PSI, the vector of observations consists of the interferometric phase double-differences and the pseudo-observations. The integer valued parameters are the ambiguities; the real-valued unknowns the topographic

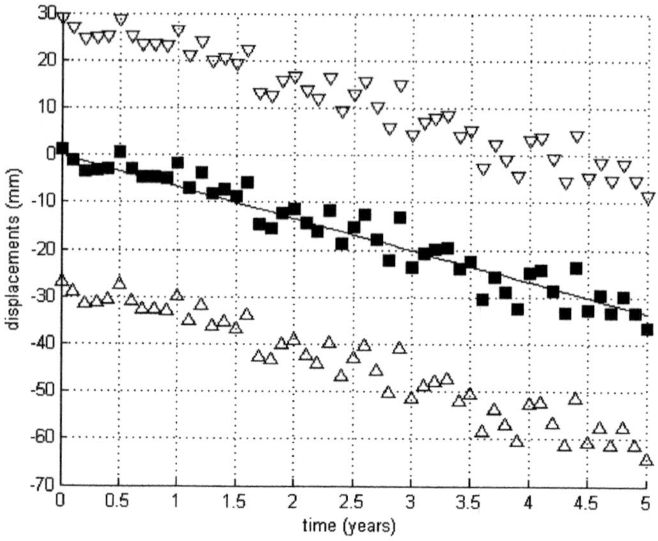

Fig. 3.11 Simulated example of the ambiguity in the displacement estimates. The squares depict the actual displacements; the triangles show the ambiguity in the displacements

heights and the deformation parameters. The solution of (3.14) is obtained in 3 steps:

1. integral float solution, obtaining the real-valued estimates $\hat{a}$ and $\hat{b}$ with the corresponding variance-covariance matrix Q,
2. mapping the float estimates $\hat{a}$ to the integer space: $\check{a} = S(\hat{a})$,
3. computation of the fixed solution of $\hat{b}$: $\check{b} = \hat{b} - Q_{\hat{b}\hat{a}} Q_{\hat{a}}^{-1}(\hat{a} - \check{a})$.

The *pull in region* of an integer vector z is denoted by S_z. It contains all real-valued ambiguity vectors that are mapped to the same integer vector z. The probability that $\hat{a}$ is mapped to z is computed by the integration of the probability density function of $\hat{a}$ over the pull-in region:

$$P(\check{a} = z) = \int_{S_z} p_{\hat{a}}(x)dx, \quad z \in Z^n, \tag{3.15}$$

where:

$$p_{\hat{a}}(x) = \frac{1}{\sqrt{\det(Q_{\hat{a}})}(2\pi)^{\frac{1}{2}n}} e^{-\frac{1}{2}\|x-a\|^2_{Q_{\hat{a}}}}. \tag{3.16}$$

The probability of correct integer ambiguity estimation $P(\check{a} = a)$ is called the *success rate*. The probability density function of the fixed solution $\check{b}$ of the real-valued parameters is a weighted sum of the integer mapping probabilities and the conditional probability density function of the float solution. Hence, it is a multi-modal distribution, not a normal distribution:

$$p_{\check{b}}(x) = \sum_{z \in Z^n} p_{\hat{b}|\hat{a}}(x|z) P(\check{a} = z). \tag{3.17}$$

Two methodologies for integer estimation are now discussed in more detail regarding computation time effectiveness and their estimation properties, i.e. integer bootstrapping and integer least-squares.

Integer bootstrapping uses a combination of integer rounding and sequential conditional least-squares adjustment techniques. The first float estimate is rounded to its nearest integer, and all subsequent float estimates are rounded taking the correlation with the previous estimate into account. The bootstrapped estimator is not unique and depends on the order of the ambiguities. In practice a decorrelating Z-transformation is applied that reduces the sequential conditional variances (Teunissen, 1995). $Q_{\hat{a}}$ is decomposed as:

$$Q_{\hat{a}} = LDL^T, \tag{3.18}$$

where L is a lower triangular matrix and D a diagonal matrix containing the conditional variances $\sigma^2_{\hat{a}_{i|I}}$. Ambiguity decorrelation increases the success rate of the bootstrapped solution $\check{a}_B$, which can be considered as a lower bound for the success rate of ambiguity resolution:

$$P(\check{a}_B = a) = \prod_{i=1}^{n} \left[2\Phi\left(\frac{1}{2\sigma_{\hat{a}_{i|I}}} \right) - 1 \right], \tag{3.19}$$

where:

$$\Phi(x) = \int_{-\infty}^{x} \frac{1}{\sqrt{2\pi}} e^{-\frac{1}{2}v^2} dv. \tag{3.20}$$

Integer Least-Squares minimizes (3.14) in the metric of the variance-covariance matrix:

$$\min_{a,b} = \|y - Aa - Bb\|^2_{Q_y}, \quad a \in Z^n, b \in R^p, \tag{3.21}$$

which can be decomposed into three orthogonal terms:

$$\|y - Aa - Bb\|^2_{Q_y} = \|\hat{e}\|^2_{Q_y} + \|\hat{a} - a\|^2_{Q_{\hat{a}}} + \|\hat{b}(a) - b\|^2_{Q_{\hat{b}|\hat{a}}}. \tag{3.22}$$

The mapping of $\hat{a}$ to the integer space is performed by minimizing the second term. Again the Z-transform is applied to minimize the search space. Although the required computation time for ILS is significantly larger than for integer bootstrapping, the ILS estimator maximizes the probability of correct integer estimation.

In this study, integer bootstrapping is applied to restrict computation time as the more optimal ILS method is time consuming. A lower bound for the success rate of ambiguity resolution is given by (3.19). An upper bound for the success rate of ambiguity resolution can be determined using the Ambiguity Dilution of Precision (ADOP). It represents the geometric mean of the conditional variances:

$$\text{ADOP} = \sqrt{\det Q_{\hat{a}}}^{\frac{1}{n}}. \tag{3.23}$$

3.4.3 Stochastic Model

The stochastic model of PSI is a superposition of contributions due to measurement noise and model imperfections (residual deformation, atmospheric signal). The variance-covariance matrix of the double-difference phase observations $\underline{y} = \underline{\varphi}_{ij}^{k}$ reads:

$$Q_y = W(Q_n + Q_{\text{defo}} + Q_{\text{atmo}})W^T, \tag{3.24}$$

where the matrix W specifies the conversion from single SAR phase observations to double-differences:

$$W = \begin{bmatrix} 1 & -1 & -1 & 1 & & & \\ \vdots & \vdots & & & \ddots & & \\ 1 & -1 & & & & -1 & 1 \end{bmatrix}. \tag{3.25}$$

The measurement noise is represented by Q_n, while Q_{defo} and Q_{atmo} represent model imperfections for respectively unmodeled deformation and atmospheric signal.

An alternative construction of the variance-covariance matrix for PSI is given by Rocca (2007) and De Zan and Rocca (2005), which is based on PS coherence measures and focuses on the temporal and geometrical (de)correlation of PS targets within the interferometric stacks.

3.4.3.1 Measurement Precision

The SLC phase precision is specified in Q_n, which is determined by measurement precision and physical PS properties; hence it varies per PS.

If the target exhibits a low phase dispersion, the probability density function of the phase observations can be approximated by a normal distribution. This has been evaluated by means of Goodness-of-Fit tests that indicate whether a dataset consisting of a number of realizations stems from a certain distribution. The Kolmogorov-Smirnov test is such a Goodness-of-Fit test. Its teststatistic is defined as the maximum absolute difference between the cumulative distribution of the dataset and the cumulative hypothesized distribution, which is in this case the normal distribution. The Lilliefors Goodness-of-Fit test (Lilliefors, 1967) defines its teststatistic similar to the Kolmogorov-Smirnov teststatistic, but also estimates the parameters of the normal distribution from the dataset. The values of the Lilliefors teststatistic have

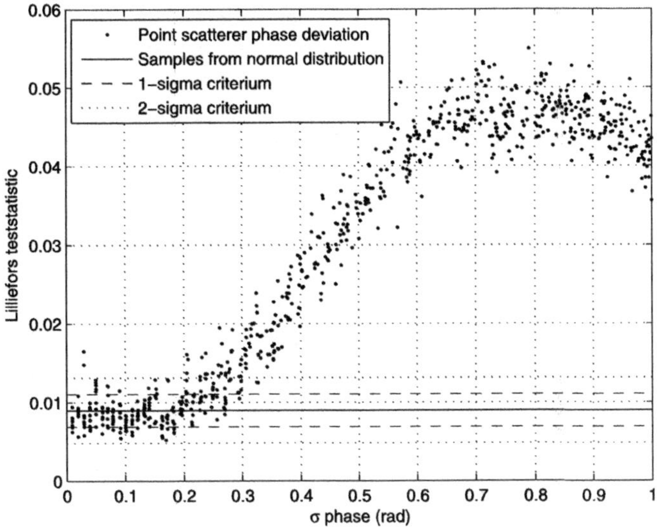

Fig. 3.12 Goodness-of-Fit teststatistics of points scatterer phase observations for the normal distribution. Only for a phase standard deviation lower than 0.3 radians, the observations can be considered normally distributed

been computed for both the point scatterer phase residuals and an equal number of independent samples from the normal distribution, see Fig. 3.12. From the deviation of the reference normal distribution, indicated by the horizontal lines, it can be concluded that the phase residuals are approximately normally distributed if the phase standard deviation is lower than 0.3 rad (~2–3 mm).

In fact, the wrapped phase observations have a multi-modal probability density distribution, that has been introduced by Adam et al. (2004):

$$\mathrm{pdf}(\psi) = \frac{\sqrt{\mathrm{SCR}} \cdot |\cos(\psi)|}{\sqrt{(\pi)}} \cdot e^{-\mathrm{SCR}\sin^2\psi}, \tag{3.26}$$

where ψ is the phase observation in a single SAR acquisition. The probability density function of the phase observations is a function of SCR (or the normalized amplitude dispersion, see (3.3)).

Another commonly used precision measure is the multi-interferogram complex coherence (Colesanti et al., 2003). However, since it can only be determined after parameter estimation, it is mixed with model imperfections and therefore not suitable for an objective description of observational precision.

3.4.3.2 Model Imperfections

Unmodeled deformation can be described in the stochastic model in Q_{defo}. PSI displacements can represent deformation due to different mechanisms. Since the knowledge on the presence and functional modeling of deformation mechanisms is often not sufficient a-priori, model imperfections can be modeled stochastically. This can be performed by means of covariance functions representing a certain spatio-temporal behavior, depending on the displacement causes, see Sect. 2.3.3.2. To apply these covariance functions in an operational way, a PS classification may be required, which will be addressed in Sect. 4.5.

Since it is complicated to describe (residual) atmospheric disturbances in a linear(ized) relation with the phase observations, they are often modeled stochastically. The spectral behavior of atmospheric signal allows to model it using a Matern class covariance function (Grebenitcharsky and Hanssen, 2005). Due to the high spatial resolution of an interferogram, it is possible to estimate the parameters of this covariance function. A simplified strategy implements an empirical covariance function that is parameterized by a variance factor and correlation length per SAR acquisition (Kampes, 2005).

Following (3.24), an example of a stochastic model for the estimation of subsidence due to hydrocarbon production could read:

$$Q_y = W\left(\sum_{p=1}^{P} \sigma_p^2 Q_p + \sigma_{\text{defo}}^2 e^{(-\frac{l_{ij}}{L_{\text{defo}}})^2} e^{(-\frac{t_{ij}}{T_{\text{defo}}})^2} + \sum_{k=1}^{K} \sigma_{k,\text{atmo}}^2 \exp^{(-l_{k,ij}^2 w^2)}\right) W^T, \tag{3.27}$$

where:

σ_p^2	variance phase observation PS p,
l_{ij}, t_{ij}	spatial and temporal difference between PS i and j,
$L_{\text{defo}}, T_{\text{defo}}$	spatial and temporal correlation length residual deformation,
σ_{defo}^2	variance residual deformation signal,
σ_{atmo}^2	variance (residual) atmospheric signal,
w	parameter related to the fractal dimension of atmospheric signal,
L_{atmo}	correlation length atmospheric signal: $w^2 = \frac{\ln(2)}{L_{\text{atmo}}^2}$.

3.4.4 DePSI Estimation Strategy

The Delft PSI (DePSI) methodology applies parameter estimation in a nested way, because of computational restrictions and the demand for observations with a high precision that can be unwrapped in the presence of atmospheric disturbances. This nested approach starts with a *first order* network of the PS candidates (PSC1) with the lowest amplitude dispersion, which implies a high likelihood of precise phase observations. The first order points form the initial (sparse) network. This network is used for initial ambiguity estimation and the initial estimation of phase screens (atmospheric and/or orbital). The first order points should be evenly distributed over the scene to result in an accurate estimation of the phase screens.

Since the system of equations is not redundant and moreover uncertainties exist in the stochastic model of the observations, observations and model errors cannot be tested. Although for P PS candidates only $(P - 1)$ independent double-differences can be formed, multiple spatial configurations are evaluated to determine which PS candidates should be accepted. These spatial configurations can have the shape of a Delaunay or 'spider' network (Kampes, 2005). The subset of accepted PS candidates in the first order network is denoted as PS1.

The *second order* PS candidates (PSC2) are selected relative to the first order network, and serve as a densification of the PS distribution. The subset of accepted second order points is denoted by PS2. To avoid that errors in the first order network will propagate undetected in the second order network, additional test procedures are applied. The densification of the PS network can be repeated to infinity.

3.5 Conclusions

The selection of Persistent Scatterer candidates is performed using the amplitude observations. In has been shown that the time and storage space consuming amplitude calibration can be replaced by a pseudo-calibration. Pseudo-calibration does not explicitly perform calibration, but adapts the normalized amplitude dispersion threshold to match the uncalibrated stack. This is done by a Monte-Carlo simulation utilizing the calibration factors that have been estimated from uncalibrated patches.

The unknown parameters (deformation, topographic heights, etc.) are estimated from double-difference phase observations. The mathematical framework of the Delft PSI estimation methodology has been introduced. Parameter estimation is first performed in a sparse first order network. After the first order network has been tested and erroneous PS candidates have been removed, further densification steps are performed.

Limitations in the PSI estimation, such as lack of redundancy and uncertainties in the stochastic model, have been pointed out. In Chap. 4 we will further analyze the effect of shortcomings in the functional and stochastic model on the estimation of the unknown (deformation) parameters. Besides accuracy of the PSI measurement technique itself, it will focus on the feasibility of PSI for the estimation of the deformation signal of interest.

Chapter 4
Quality Control

In Chap. 3, the mathematical framework for the estimation of earth surface deformation has been explained. The application of PSI for deformation monitoring is nowadays moving towards deformation phenomena that have a small magnitude in areas that suffer from temporal decorrelation. Hence, the precision and the reliability of the deformation estimates become increasingly important. Therefore, this chapter is dedicated to quality control.

Regardless of the measurement technique, quality control for deformation monitoring has intrinsically two components:

- precision and reliability of the estimated parameters, and
- the relation of the estimated parameters to the signal of interest.

For subsidence monitoring due to gas extraction in the Netherlands both aspects are equally important. Related to the first component, a quantification of precision and reliability of PSI is required to evaluate the potential of the technique in comparison to the leveling measurements. While the deformation signal of interest is subsidence due to gas extraction, InSAR observes movements of scatterers regardless of the deformation mechanism. The challenge is therefore to discriminate subsidence due to gas extraction in the possible presence of multiple deformation causes (shallow compaction, instability of buildings, etc.).

Chapter 4 consists of two parts. The Sects. 4.1 to 4.4 will address the uncertainties in the functional and stochastic model. The effect of possible model errors on the deformation estimates is evaluated by means of the mathematical framework of Chap. 3. The precision of PSI observations is assessed empirically by means of the Delft corner reflector experiment. Section 4.5 investigates the information that can be utilized to increase the idealization precision for the estimation of the deformation signal of interest. This section focuses further on the characterization of Persistent Scatterers and the use of a-priori knowledge on the spatio-temporal behavior of the signal of interest.

4.1 Precision and Reliability in PSI

Precision is the dispersion of a stochastic variable around its mean. In this book, the precision refers to the square root of the variance, i.e. the standard deviation (1-sigma criterium) of a stochastic variable, if not indicated otherwise. In PSI, the precision of an observation is determined by both the measurement precision and the physical properties of the measurement target. These two components of the precision cannot easily be separated. Hence, it is complicated to construct the variance-covariance matrix of InSAR observations. Therefore, a validation of the stochastic model of InSAR has been performed in a controlled corner reflector experiment using independent leveling measurements, see Sect. 4.4.

V.B.H. (Gini) Ketelaar, *Satellite Radar Interferometry*,
Remote Sensing and Digital Image Processing 14,

Reliability is defined as the sensitivity for and detectability of model imperfections (Teunissen, 2000b). In the PSI system of observation equations (see (3.11)), there is no redundancy. Hence, testing on model errors and outliers in the observations is not possible. However, if the success rate of ambiguity resolution equals 1, the ambiguities can be considered to be deterministic. As a result, the PSI system of observation equations becomes redundant and geodetic testing techniques can be applied to assess the reliability (ibid.). In the following paragraphs, the effect of uncertainties in the functional and stochastic model on the parameter estimates are evaluated based on the assumption of correct phase unwrapping. It gives an indication of sensitivity of the deformation estimates for errors in the different components in the mathematical framework. However, it is stressed that this is the most optimistic scenario, since correct phase unwrapping has been assumed.

If the success rates of phase unwrapping can not be assumed equal to 1—which is not unlikely in rural areas—reliability cannot be assessed in the mathematical framework of PSI estimation. However, in this book it will be shown that by utilizing multiple independent satellite tracks that observe the same deformation signal, redundancy is introduced and hence a statement about the reliability of the deformation estimates can be made. This method is referred to as multi-track PSI and will be discussed in Chap. 5.

4.2 Influence of Imperfections in the Functional Model

In this section, the influence of shortcomings in the functional model of (3.11) on the unknown parameters is investigated. An example is the possible bias in the PS height and displacement rate estimates that is caused by an inaccurate sub-pixel position. The effect of model errors on the unknown parameters can be determined as:

$$\nabla\hat{x} = (A^T Q_y^{-1} A)^{-1} A^T Q_y^{-1} \nabla y, \tag{4.1}$$

where A specifies the functional relation from (3.11), Q_y is the variance-covariance matrix of the double-differences, ∇y is the model error, and $\nabla\hat{x}$ its influence on the parameter estimates (Teunissen, 2000b).

The unknown parameters of interest in deformation monitoring are the displacement estimates. For a further characterization of the PS targets, the topographic height estimates are valuable as well. From the topographic height it can for example be determined if the PS reflection stems from a roof top or ground level. Hence, the effect of model errors on the displacement and topographic height estimates is addressed in this section. The potential model errors that are investigated are inaccurate sub-pixel positions, sidelobe observations, and orbital inaccuracies. Furthermore, the phase unwrapping success rates are addressed.

All results in this section are based on simulations. To obtain a realistic acquisition geometry (distribution of perpendicular and temporal baselines), samples have been taken from the configurations of six existing ERS tracks that cover the northern part of the Netherlands. Furthermore, it has been chosen to model deformation as a constant displacement rate (velocity). In this way, the redundancy in the functional model is optimally utilized. Moreover, subsidence above gas fields that are in production during the entire monitoring period can be unwrapped in time by using a model of linear displacements (see Sect. 3.4.2, Fig. 3.11).

The success rate of phase unwrapping in the evaluation of model errors is considered to be 1. This implies that a quantification of the sensitivity of deformation and height estimates for different model errors is obtained in the most optimistic scenario. In Sect. 4.2.4 it will be shown that success rates of 1 cannot be guaranteed in areas with low PS densities. It has already been pointed out in the introduction of this chapter that additional methods are required to assess the reliability of the results. In this research, multi-track PSI has been developed as a methodology for reliability assessment and will be discussed separately in Chap. 5.

4.2.1 Sub-pixel Position

Inaccuracies in the sub-pixel position cause inaccuracies in the PS height and velocity estimates (Perissin, 2006). The phase contribution due to the sub-pixel position $\varphi_{\text{obj},ij}$ can be deduced from (3.11):

$$\varphi^k_{\text{obj},ij} = \frac{4\pi}{\lambda}\frac{B^\perp_i}{R^m_i \tan\theta^m_i}\eta^m_{ij} + \frac{2\pi}{v}(f^m_{\text{dc},i} - f^k_{\text{dc},i})\xi^m_{ij}. \tag{4.2}$$

It is a function of the slant-range (η) and azimuth (ξ) sub-pixel coordinates. The model error ∇y is the error that is introduced by errors in the range and azimuth sub-pixel position $\Delta\eta^m_{ij}$ and $\Delta\xi^m_{ij}$. Model error samples have been computed using the acquisition geometries of six ERS tracks in the northern part of the Netherlands. Subsequently, the effect on the PS height and velocity estimates for a single arc have been computed using (4.1). The correlation between the double-differences has been taken into account in Q_y, see Sect. 3.4.3. The design matrix contains the functional relation between the phase observations and the unknown PS height and velocity:

$$A = -\frac{4\pi}{\lambda}\begin{bmatrix} \frac{B^\perp_i}{R^m_i \sin\theta^m_i} & T^k \end{bmatrix}, \tag{4.3}$$

where T^k is the temporal baseline that relates to the unknown PS velocity.

From Fig. 4.1 it can be deduced that the effect of errors in the range sub-pixel position results in errors of ~0–4 meters on the PS heights, while they do not affect the PS velocities (a standard oversampling factor of 2 is assumed). The error in the PS height estimates directly affects the horizontal location of the PS (see Sect. 4.5.2).

Errors in the azimuth sub-pixel position lead to small errors (<0.25 m) in the PS heights (again assuming an oversampling factor of 2). The velocities are affected ~0–0.5 mm/year. The effect of these systematic errors increase when images with a higher Doppler deviation than 500 Hz with respect to the master are included: up to 3 mm/year for the PS velocities.

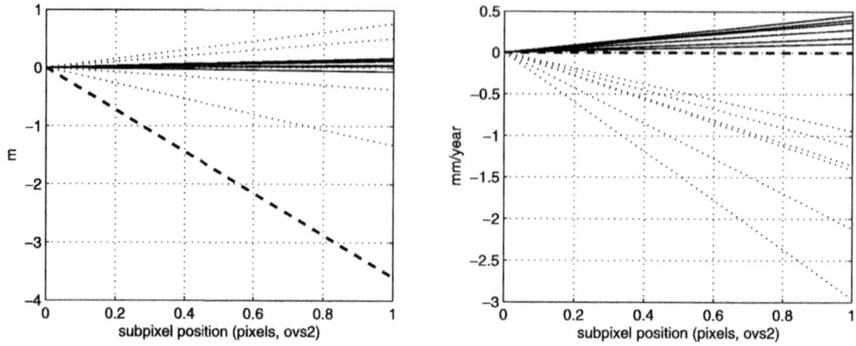

Fig. 4.1 The effect of sub-pixel position errors in range (dashed) and azimuth (solid) on PS height (left) and velocity (right) estimates (oversampling factor 2). The dotted lines represent the situation where acquisitions with high Doppler deviations (>500 Hz) have been included

In the implementation of the DePSI concept for monitoring subsidence due to gas extraction, the sub-pixel position has not been incorporated in the functional model, see Chap. 6. The reason behind this is to avoid a decrease in redundancy for parameter estimation. Moreover, there is a correlation between range sub-pixel position and the PS heights (see Sect. 3.4), which complicates their joint estimation. If all acquisitions in a PSI stack are oversampled by a factor 2 and if scenes with high Doppler deviations are excluded, PS velocity errors of maximum ~0.5 mm/year can be expected for particular scatterers. However, when including high Doppler images for subsidence monitoring in areas like Groningen (displacement rates <1 cm/year), the estimation of the azimuth sub-pixel position should be included in the functional model.

4.2.2 Sidelobe Observations

The spatial signature of PS targets that behave as distinct point scatterers is a sinc pattern (Oppenheim et al., 1983; Cumming and Wong, 2005). The sinc function reads:

$$\mathrm{sinc}(x) = \frac{\sin(\pi x)}{\pi x}. \tag{4.4}$$

The sidelobes of this sinc pattern are located in other resolution cells than the main lobe of the scatterer. The phase in every second sidelobe is equal to the phase in the main lobe, while phases in the interleaved sidelobes are π radians reversed, see Fig. 4.2. As the phase observations of all sidelobes refer to the same physical target, they are not independent. Since the phase behavior of sidelobes is coherent, there is a likelihood that they are detected as PS candidates. The effect of the different range and azimuth location compared to the PS they refer to is investigated in this section.

The influence of sidelobe observations on the PS height and velocity estimates is assessed. The model errors in the sidelobe observations comprise of:

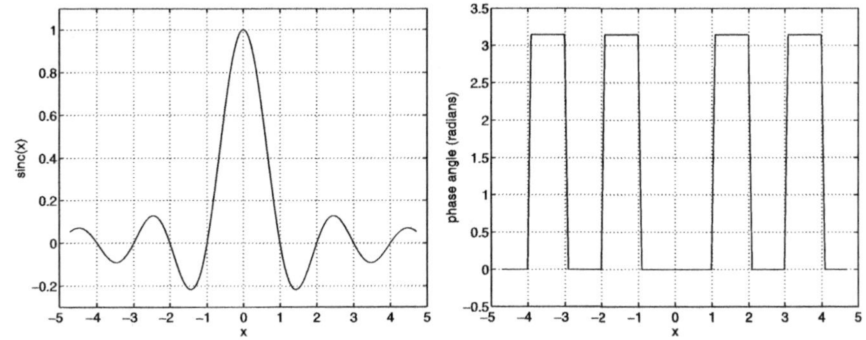

Fig. 4.2 Sinc function (left) and the corresponding phase angles (right)

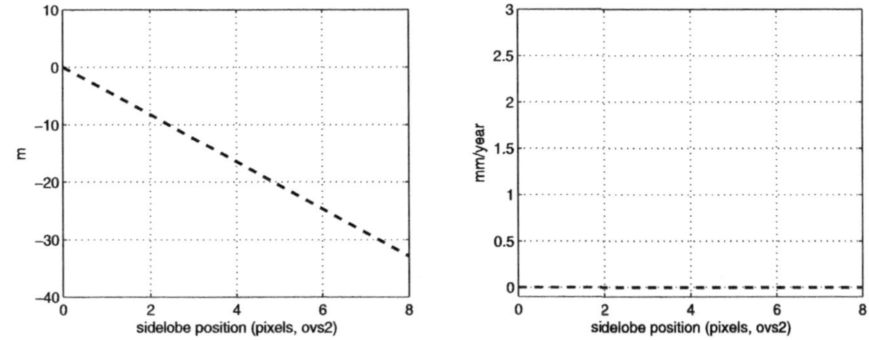

Fig. 4.3 Influence of sidelobe observations on the height (left) and velocity (right) estimates. The heights are significantly affected, whereas the velocity estimates of the sidelobes are the same as the estimates for the PS they refer to

- errors in $B_i^{\perp}$, R_i^m, and θ_i^m, and
- errors in the reference phase $\frac{4\pi}{\lambda} B \sin(\theta_i^m - \alpha)$.

Because the baseline B and its orientation α depend on the satellite positions only, they have been not been changed for the sidelobe observations. The variation in incidence angle was computed by means of the flat earth approximation:

$$\theta_i^m = \arccos\left(\frac{H_{\text{sat}}}{R_i^m}\right); \quad \frac{d\theta_i^m}{dR} = \frac{H_{\text{sat}}}{(R_i^m)^2\sqrt{1-(H_{\text{sat}}/R_i^m)^2}}, \tag{4.5}$$

where H_{sat} is the satellite height.

From Fig. 4.3 it can be concluded that the PS height estimates are significantly affected: almost 20 meters at a four pixels' range distance from the target location. The velocity estimates are equal to the velocity estimates of sidelobes' PS target. Hence, the estimation of the displacement rates is not biased by including sidelobe observations. However, it has to be realized that sidelobe observations are duplicates

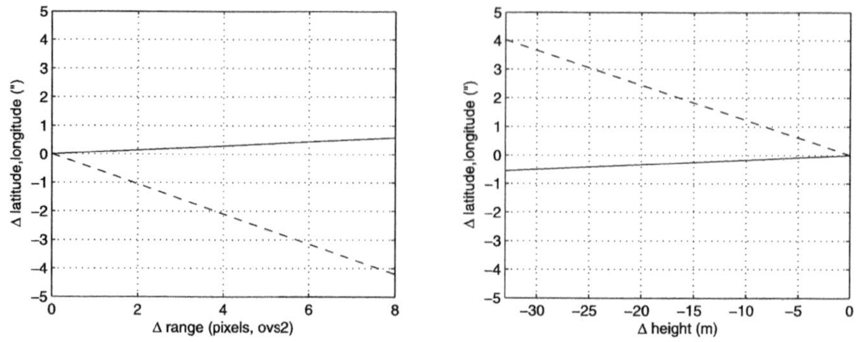

Fig. 4.4 The effect of range coordinate and height deviations on the geolocation: latitude (solid) and longitude (dashed). The difference in geographic coordinates is in the order of several seconds (1" is ~30 meters)

of the phase observation of one physical scatterer. They are not independent and do not contribute to the redundancy in the estimation of the deformation signal of interest.

Due to the impact on the height estimates, the geocoding of sidelobes that are incorrectly detected as PS is erroneous. The range coordinate and the height of the sidelobe have been varied according to the ratio depicted in Fig 4.3. From Fig. 4.4 it can be deduced that the geographic coordinates change in the same way for an incorrect range coordinate as for an incorrect height estimate. Consequently, the geolocation of a sidelobe would coincide with the geolocation of the actual PS target if its height estimate compensates for the different range location. From Fig. 4.3, it follows that a negative change in range coordinate with respect to the position of the main lobe (the actual target) results in a positive change of the sidelobe's height estimate. Therefore, the effect on the geocoding of the sidelobe's height estimate is canceled by its range coordinate. Hence, the geographic position of the sidelobe will be exactly the position of the PS it refers to. This will result in several estimates plotted on top of eachother in a point-wise visualization.

4.2.3 Orbital Inaccuracies

Due to the wide spatial extent of the deformation signal of interest, the influence of residual orbital components has to be investigated. Orbit errors can be decomposed into errors in along-track, across-track and radial direction. Since along-track errors are sufficiently corrected for in the coregistration step, the analysis of systematic phase errors restricts to the across-track and radial direction (Hanssen, 2001). The effect of orbit errors is the incorrect estimation of the reference phase that will propagate systematically from near to far range, and errors in the computation of the perpendicular baseline.

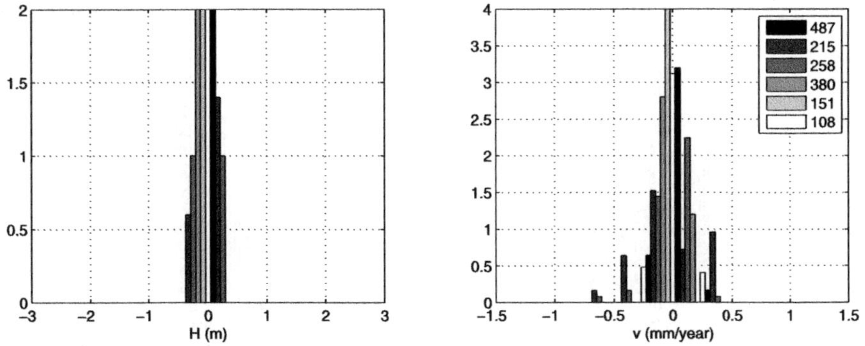

Fig. 4.5 Uncertainty in PS height and velocity estimates between near and far range (100 km) due to random across-track orbit errors (σ 8 cm) for six tracks. For each simulation of random across-track errors per track, the corresponding phase observation errors have been computed, which are induced by errors in the reference phase between near and far range. Subsequently, the effect on the velocity and height estimates have been computed. The histograms show the results of 50 simulations per track

To investigate the effect of orbit errors, the effect of residual across-track and radial errors has been investigated for six existing ERS tracks. The simulation consists of the following steps:

1. generation of random across-track and radial errors with a standard deviation of respectively 8 and 5 cm for all acquisitions,
2. adaptation of the baseline sizes and orientations,
3. computation of the model errors induced in the reference phase between near and far range, and
4. adaptation of the design matrix with the adapted perpendicular baselines.

Figures 4.5 and 4.6 show the effect of random orbit errors on the velocity and height estimates. Radial orbital errors have the largest effect on the velocity estimates: up to 1 mm/year. Across-track errors result in maximum velocity deviations of 0.5 mm/year. Deviations in the height estimates due to radial orbit errors are up to 2 meters, whereas the height deviations are less than 0.5 meter for across-track errors. Regarding the velocity estimates for areas with a large spatial extent, this implies that a small spatial trend can be present. The presence and elimination of spatial trends will be further discussed in the multi-track approach in Chap. 5.

4.2.4 Phase Unwrapping in the Presence of Atmospheric Disturbances

For the application of PSI in rural areas with a relatively low PS density, the estimation of atmospheric signal plays an important role in quality assessment. In the first order network of PS candidates, the atmospheric disturbance is increasing with arc

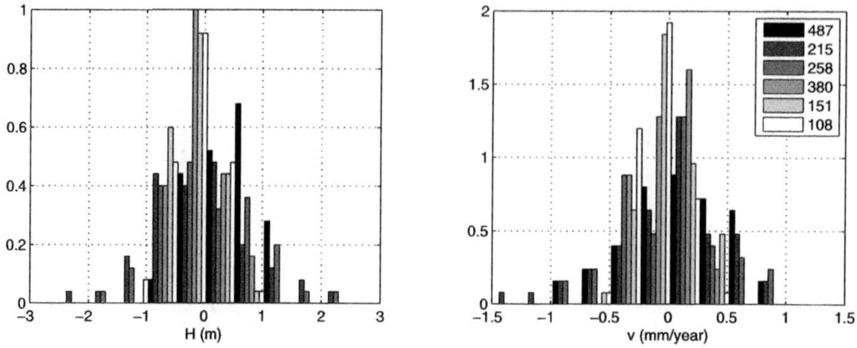

Fig. 4.6 Uncertainty in PS height and velocity estimates between near and far range (100 km) due to random radial orbit errors (σ 5 cm) for six tracks. For each simulation of random radial errors per track, the corresponding phase observation errors have been computed, which are induced by errors in the reference phase between near and far range. Subsequently, the effect on the velocity and height estimates have been computed. The histograms show the results of 50 simulations per track

length and therefore decreasing the success rate of correctly unwrapping the fractional phase difference observations. After temporal unwrapping per arc, a spatial network check is performed tracing and removing PS candidates that are responsible for spatial unwrapping misclosures. However, this procedure does not guarantee protection against type II errors: PS candidates that are not rejected even though the spatial unwrapping is erroneous. After estimation and subtraction of the atmospheric phase screens (APS) for every acquisition, the unwrapping success rate increases. Optimization of the unwrapping success rates is dependent on:

- the measurement precision of the interferometric phase differences,
- the magnitude (power) of the atmospheric regimes (dependent on the weather conditions),
- the PS density: the spatial sampling of the APS, and
- the number of acquisitions: the master APS can be estimated with a higher accuracy when more acquisitions are available.

This section investigates the dependence of the success rates on observational precision and PS density in the presence of atmospheric disturbances. The success rates of integer bootstrapping (see (3.19)) have been used, which serve as a lower bound (Teunissen, 2001b). The number of interferograms in a stack has been set to 25, which serves as a minimum for PSI (Colesanti et al., 2003). Thus, the resulting success rates will be a lower bound regarding the number of acquisitions for APS estimation.

A series of APS have been simulated, consisting of scaled regimes with their fractal dimension (Hanssen, 2001). Three regimes are distinguished: regime I that covers large scale variations, regime II that covers scales from resolution level to the thickness of the turbulent layer, and regime III representing small scale variability. Since regime III is probably not caused by atmosphere, it has been excluded

Table 4.1 Bootstrapping success rates for phase unwrapping in stacks of 25 interferograms with simulated phase observations and atmospheric phase screens. The percentage of arcs with a success rate higher than 0.5, 0.8 and 0.99 respectively is listed. This table shows the influence of observational precision and the PS density on the success rates. The observational precision is expressed in cycles: 1/20 cycle corresponds with a precision of 2.8 mm, $1/10$ cycle corresponds with a precision of 5.6 mm

acquisitions	σ_ϕ (cycles)	before APS estimation success rate			after APS estimation success rate		
		>0.5	>0.8	>0.99	>0.5	>0.8	>0.99
100 PS/km^2							
25	1/20	100 (%)	100	99	100 (%)	100	100
25	1/10	71	34	18	75	40	21
5 PS/km^2							
25	1/20	100 (%)	98	95	100 (%)	100	100
25	1/10	38	8	3	40	20	7

from the simulations. The transition between regimes I and II has been set at 2 kilometers distance. Regimes I and II have a power law exponent of $-5/3$ and $-8/3$ respectively.

Table 4.1 shows that high success rates can be achieved at a lower PS density, provided that the observational precision is high. For low PS densities, the increase in success rate before and after APS estimation is more significant, since arcs lengths are larger and will benefit more from APS removal. Compared to success rates in areas with an urban PS density, similar success rates can be achieved in rural areas, provided that arcs exist with a phase dispersion lower than $1/20$ cycle (<3 mm). However, Fig. 6.20 in Chap. 6 shows that this requirement on the precision is not necessarily met. Table 4.1 also shows that success rates of 1 cannot be guaranteed, especially not in areas with a PS density of $\sim$5 PS/km^2. This implies that the ambiguities cannot considered to be deterministic, therefore a direct reliability assessment using the functional model of (3.11) is not possible due the lack of redundancy. In this situation other tools for reliability assessment have to be applied, such as multi-track PSI, see Chap. 5.

4.3 Imperfections in the Stochastic Model

This section gives an overview of the uncertainties in the components of the stochastic model: measurement noise and model imperfections. Furthermore, it introduces a precision parameterization of the variance-covariance matrix that is independent of the spatial and temporal reference, i.e. reference PS and master acquisition.

4.3.1 Measurement Precision

Contrary to other geodetic measurement techniques, the stochastic model of the PSI observables is poorly known a-priori. This is caused by the dependence of the measurement precision on the physical PS properties. Due to the uncertainty in geocoding, it is complicated to identify the origin of the reflection. Moreover, the reflection pattern of surrounding scatterers may interfere, which introduces additional noise in the phase observations.

Section 3.2.1 has shown that the phase variance is a function of the Signal-to-Clutter Ratio, which is directly related to the normalized amplitude dispersion, see (3.3). Phase variances as a function of SCR can be used as a-priori variances indicating the relative weights of different PS. This would imply only one unknown variance factor (a scaling factor): $Q_n = \sigma^2 Q$, where Q represents the a-priori variance-covariance matrix. This procedure has been followed in the Delft corner reflector experiment, which will be separately explained in Sect. 4.4.

Another option is to estimate a variance factor for each PS p: $\sum_{p=1}^{P} \sigma_p^2 Q_p$. This implies that significantly more variance factors have to be estimated while the redundancy in the system of equations remains unchanged. As a result, the estimation of the variance factor per PS will be less precise. The possibilities and limitations of variance component estimation will be addressed in Sect. 4.3.3.

4.3.2 Separation of Unmodeled Deformation and Atmospheric Signal

As knowledge about the functional model of the deformation signal is generally limited, a spatio-temporal filtering approach is used to address the phase residuals to unmodeled (non-linear) deformation and atmospheric signal. Based on the assumption that atmospheric signal is uncorrelated in time whereas unmodeled deformation is not, the residuals are separated (Ferretti et al., 2001).

If the functional model of the deformation signal is correct, all residuals should be addressed to atmospheric signal (and measurement noise). The presence of unmodeled deformation signal is however not known beforehand. To preserve all possible unmodeled deformation regimes, all residuals are addressed to deformation signal. As a result, the estimated time series appears noisier and seems to be of lower quality. However, when addressing high variability in the observations to atmospheric signal followed by subsequent subtraction, there is a risk of subtracting unmodeled deformation as well. This implies a loss of the signal of interest and hence unrecoverable model errors.

In the DePSI estimation procedure (see Sect. 6.1.3), the phase residuals are separated into atmospheric signal and unmodeled deformation by the application of a moving average filter in time that has the characteristics of a low pass filter (ibid.). The larger the window size of the moving average filter, the more high frequency

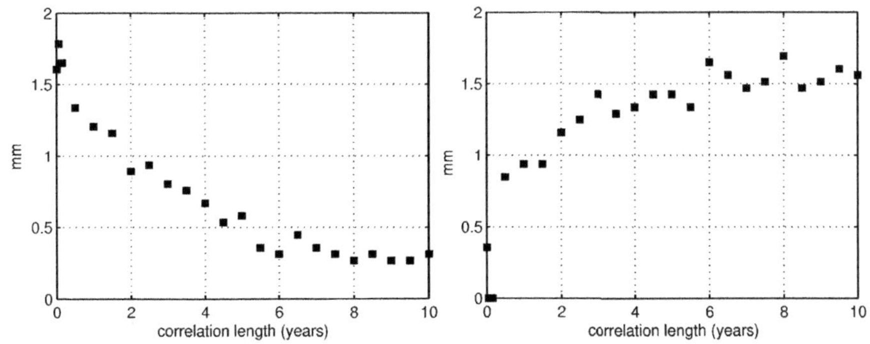

Fig. 4.7 Separation of unmodeled deformation (left) and atmospheric signal (right). The correlation length of the deformation signal corresponds with the window size of the temporal moving average filter. For small values of the correlation length, phase residuals are identified as unmodeled deformation. The larger the size of the correlation length of the deformation signal, the more residual components are identified as atmospheric signal

signals with be removed from the phase residuals. Thus, the risk of incorrectly identifying unmodeled deformation as atmospheric signal increases with the window size of the moving average filter.

Figure 4.7 shows a simulated example of the division of the residuals in time into atmospheric signal and unmodeled deformation. The simulated stack consists of 30 acquisitions that span a 15 years' period. The simulated acquisitions only contain atmospheric signal. Unmodeled deformation is absent, as well as measurement noise. Hence, all residuals should be addressed to atmospheric signal. The size of the temporal moving average filter is expressed as the correlation length of the deformation signal. A large correlation length corresponds with the absence of unmodeled deformation: all high frequency signals in the phase residuals can be filtered out and identified as atmospheric signal. From Fig. 4.7 it can be deduced that for temporal correlation lengths above 6 years, indeed the residuals are attributed to atmospheric signal. Furthermore, it can be seen that part of the atmospheric signal will be incorrectly interpreted as deformation signal if the correlation length is set to values lower than 6 years.

The separation of non-linear deformation and atmospheric signal is a trade-off between preservation of possible unmodeled deformation and precision of the displacement estimates. Hence, a high precision of the displacement estimates can be at the expense of the leaking of unmodeled deformation into the atmospheric signal.

In stead of filtering unmodeled deformation by means of a low pass filter, it is recommended to incorporate both atmospheric signal and unmodeled deformation in the stochastic model of the PSI system of equations. This would integrate the uncertainties in the quality description. The next section addresses possibilities and limitations of the estimation of stochastic parameters in the PSI estimation procedure.

4.3.3 Possibilities and Limitations of Variance Component Estimation

Variance component estimation (VCE) (Teunissen, 1988) performs the estimation of stochastically modeled parameters. Examples of stochastic model parameters are the variance factor for the measurement noise and the spatial correlation length of atmospheric disturbances. Three things have to be kept in mind when considering the application of VCE:

1. VCE can only be applied if there is redundancy in the mathematical model,
2. only independent stochastic model parameters can be estimated, i.e. there is no rank deficiency in the VCE system of equations, and
3. the precision of the estimation of stochastic model parameters decreases with the number of stochastic parameters to be estimated.

In Sect. 2.3.1 it is pointed out that the variance components are computed using the adjustment residuals. Hence, redundancy is required. As a result, VCE can only be applied in PSI if the ambiguities are considered deterministic. Another condition for the application of VCE is the independence of the variance components. This implies that model imperfections that are modeled with the same covariance function but with different magnitudes and correlation lengths cannot be estimated independently.

The elementary stochastic modeling of residual signal is by means of an (empirical) covariance function as a function of variance, spatial and/or temporal correlation length or power (if model deviations are unbounded). The usage of more advanced covariance functions depends on the a-priori knowledge of the signal. For atmospheric disturbances, which are spatially but not temporally correlated, more advanced covariance functions are described by Grebenitcharsky and Hanssen (2005).

If the conditions on redundancy and the independence of the variance components are fulfilled, the achievable precision of the variance component estimates has to be evaluated. The more variance components are estimated, the less precise they will be. Additionally, the correlation between different variance components has to be considered. This correlation determines the possible leakage of one stochastically modeled signal into the other.

In the VCE procedure that has been described in Sect. 2.3.1, only variance factors have been considered as variance components. Q_y could be decomposed as a linear combination of variance factors and cofactor matrices. For spatio-temporally correlated signal however, the relation between the cofactor matrices Q and (some of) the variance components is usually non-linear. Therefore, the variance-covariance matrix of the observations is developed in a Taylor series. For the stochastic model of PSI observations, consisting of a superposition of measurement noise and noise due to atmospheric signal and unmodeled deformation (see (3.24)), this decomposition would read:

$$Q_y = Q_0 + W\Bigg(\frac{\delta Q_n}{\delta\sigma_n^2}d\sigma_n^2 + \frac{\delta Q_{\mathrm{defo}}}{\delta\sigma_{\mathrm{defo}}^2}d\sigma_{\mathrm{defo}}^2 + \frac{\delta Q_{\mathrm{defo}}}{\delta L_{\mathrm{defo}}}dL_{\mathrm{defo}} + \frac{\delta Q_{\mathrm{defo}}}{\delta T_{\mathrm{defo}}}dT_{\mathrm{defo}}$$

$$+ \frac{\delta Q_{\mathrm{atmo}}}{\delta\sigma_{\mathrm{atmo}}^2}d\sigma_{\mathrm{atmo}}^2 + \frac{\delta Q_{\mathrm{atmo}}}{\delta L_{\mathrm{atmo}}}dL_{\mathrm{atmo}}\Bigg)W^T, \tag{4.6}$$

where the matrix W converts the phase observations from single acquisitions to spatio-temporal double-differences. In total, six variance components are estimated using (4.6):

- one variance factor for the measurement precision (σ_n^2),
- one variance factor, spatial correlation length and temporal correlation length for unmodeled deformation signal (σ_{defo}^2, L_{defo}, and T_{defo}), and
- one variance factor and spatial correlation length of unmodeled atmospheric signal (σ_{atmo}^2 and L_{atmo}).

The variance components are obtained in an iterative way. Due to the decomposition of Q_y into a known part (Q_0) and an unknown part, the VCE system of equations differs slightly from Sect. 2.3.1:

$$l_k = \underline{\hat{e}}^T Q_y^{-1} Q_k Q_y^{-1} \underline{\hat{e}} - \mathrm{tr}(Q_y^{-1} P_A^{\perp} Q_k). \tag{4.7}$$

To illustrate the dependency of the variance component precision (see (2.14)) on the redundancy and the number of stochastic parameters, consider the following simulated example of residual deformation signal. The simulated residual deformation signal only contains measurement noise and spatially correlated noise:

$$Q_y = W(\sigma_n^2 I + \sigma_{\mathrm{defo}}^2 e^{(-\frac{Q_l}{L}^2)})W^T, \tag{4.8}$$

and its Taylor expansion:

$$Q_y = Q_0 + WW^T d\sigma_n^2 + W(e^{(-\frac{Q_l}{L}^2)})W^T d\sigma_{\mathrm{defo}}^2 + 2\sigma_{\mathrm{defo}}^2 W\left(\frac{Q_l^2}{L^3}e^{(-\frac{Q_l}{L}^2)}\right)W^T dL, \tag{4.9}$$

where Q_l is the PS distance matrix. The observations are distributed over an area of 5 × 5 km. Their expectation value is 0, their correlation length is 2.5 km, and the variance of measurement and model noise is 1 and 4 mm^2 respectively. The observations have been simulated using the Cholesky decomposition $Q_y = R^T R$. Based on this decomposition, the observations can be computed as $\underline{y} = R^T \underline{n}$, with $\underline{n}$ a vector of standard normal distributed variables.

Figure 4.9 shows the variance component estimates divided by their standard deviation ($\underline{\hat{\sigma}}/\sigma_{\hat{\sigma}}$) for the joint estimation of three variance components and for the estimation of the measurement variance factor only. It is clear that the precision of the variance component estimates increases with increasing redundancy and that the precision of the variance component estimates is higher when less variance components have to be estimated. Furthermore, it can be noticed that the model noise component is not very well estimable; the standard deviation of its estimation is approximately equal to model noise variance itself.

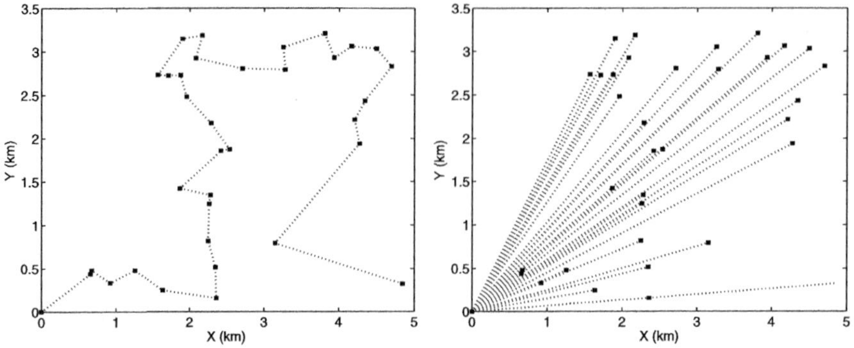

Fig. 4.8 Network configurations: (left) shortest distance and (right) star network. The observations are the spatial differences

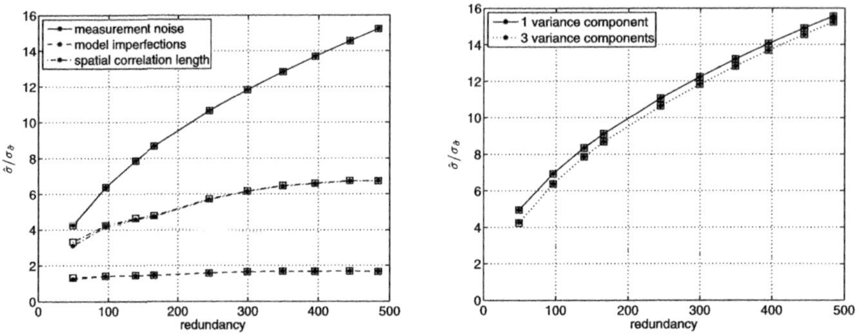

Fig. 4.9 Left: variance component estimate ($\hat{\underline{\sigma}}$) divided by its standard deviation ($\sigma_{\hat{\sigma}}$) for the estimation of both measurement and spatially correlated noise. Right: variance component estimate ($\hat{\underline{\sigma}}$) divided by its standard deviation ($\sigma_{\hat{\sigma}}$) for the estimation of measurement noise only. The precision of the variance components increases if less components are included in the estimation. The stars and squares represent variance component estimates obtained from network designs that are based on different spatial combinations of the same observations: star and path network, see Fig. 4.8. The precision of the variance components is independent of the network type

Finally, the influence of the network configuration on the precision of VCE has been investigated. Hence, simulations have been performed both for a star network and a 'shortest distance' network, see Fig. 4.8. The results for both network configurations are depicted in Fig. 4.9: the precision of the estimated variance components is the same. This can be explained from the fact that the linear combinations in both network configurations have been constructed from the same observations.

4.3.4 Dilution of Precision

The stochastic model in the PSI estimation procedure is represented by the variance-covariance matrix. Regardless if VCE is applied or not, there is a need of a quality measure for the variance-covariance matrix of both the observations and the parameter estimates. Due to the correlation induced by the double-differences and spatio-temporal correlation of model imperfections, the variance-covariance matrices of both the observations and the parameter estimates are full matrices. Restricting the precision assessment to the variances on the main diagonal results in a precision measure that is dependent on the spatio-temporal reference for the double-differences, see Sect. 3.3.2. However, double-differences that are created using different spatio-temporal references originate from the same set of phase observations. Hence, the information in their variance-covariance matrix is the same.

In this section, it is shown that a scalar precision measure exists that describes the quality of the variance-covariance matrix independent of the spatio-temporal reference: the Dilution of Precision. In analogy with the theory of ambiguity resolution for GPS, where it is shown that the Ambiguity Dilution of Precision (ADOP) (Teunissen and Odijk, 1997) is independent of the chosen reference satellite, the independence of PSI double-differences on the chosen spatio-temporal reference will be demonstrated. It is based on the theory on admissible ambiguity transformations (ibid.), which states that transformations to other references are admissible if their determinant equals ± 1. This follows from the property that the transformation matrices and their inverses only contain integer entries.

Consider a set of P PS that are observed in K acquisitions. The double-difference phase observations relative to reference PS 1 and acquisition time $k=1$ are denoted as φ_{p1}^{k1} for $p=2\dots P$ and $k=2\dots K$. The double-differences can be transformed to reference PS 2 by means of the transformation matrix T:

$$\underline{\varphi}_{p2}^{k1} = T\underline{\varphi}_{p1}^{k1}. \tag{4.10}$$

The new variance-covariance matrix is computed by applying the propagation law:

$$Q_{\varphi_{p2}^{k1}} = T Q_{\varphi_{p1}^{k1}} T^T. \tag{4.11}$$

The transformation matrix T has the following shape for P PS:

$$\begin{bmatrix} \varphi_{12}^{k1} \\ \varphi_{32}^{k1} \\ \vdots \\ \varphi_{P2}^{k1} \end{bmatrix} = \begin{bmatrix} -1 & 0 & \dots & 0 \\ -1 & 1 & \dots & 0 \\ \vdots & \vdots & \ddots & \vdots \\ -1 & 0 & \dots & 1 \end{bmatrix} \begin{bmatrix} \varphi_{21}^{k1} \\ \varphi_{31}^{k1} \\ \vdots \\ \varphi_{P1}^{k1} \end{bmatrix}. \tag{4.12}$$

Similarly, the transformation of the PSI double-differences from master acquisition time $k=1$ to $k=2$ reads:

$$\begin{bmatrix} \varphi_{p1}^{12} \\ \varphi_{p1}^{32} \\ \vdots \\ \varphi_{p1}^{K2} \end{bmatrix} = \begin{bmatrix} -1 & 0 & \dots & 0 \\ -1 & 1 & \dots & 0 \\ \vdots & \vdots & \ddots & \vdots \\ -1 & 0 & \dots & 1 \end{bmatrix} \begin{bmatrix} \varphi_{p1}^{21} \\ \varphi_{p1}^{31} \\ \vdots \\ \varphi_{p1}^{K1} \end{bmatrix}. \tag{4.13}$$

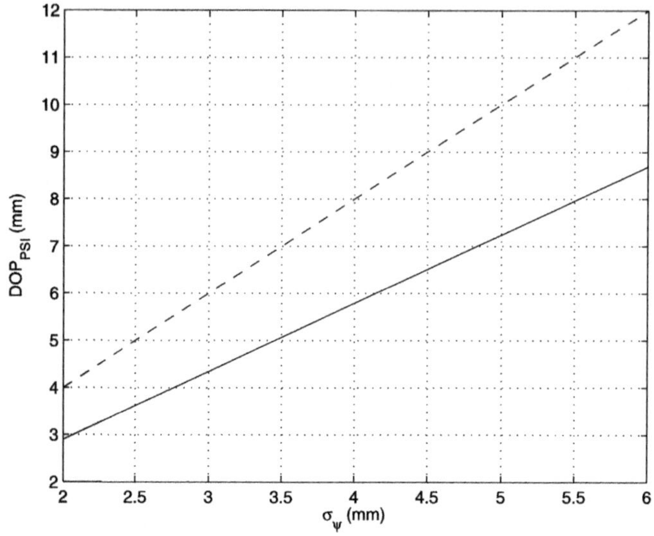

Fig. 4.10 $\mathrm{DOP_{PSI}}$ for PSI double-difference phase observations (solid) as a function of standard deviation of a phase observation in a single SAR acquisition (expressed in mm). The dashed line depicts $\mathrm{DOP_{PSI}}$ if the double-difference phase observations would have been uncorrelated (the geometric mean). The DOP value decreases with a lower standard deviation of the observations (higher precision), both for correlated double-difference observations and the geometric mean. The DOP value is a scalar that represents the precision of the variance-covariance matrix. This precision is independent of the spatio-temporal reference. However, it is one specific function of the variance-covariance matrix. A complete comparison of two variance-covariance matrices is performed through the generalized eigenvalue problem, see (7.11), which indicates the precision contained in the variance-covariance matrix for different eigenvectors

The transformation matrices and their inverses have all integer entries, and their determinant equals 1. Therefore the precision measure defined by the determinant is invariant for the PSI reference system:

$$\det Q_{\varphi^{k1}_{p2}} = (\det T)^2 \det Q_{\varphi^{k1}_{p1}} = \det Q_{\varphi^{k1}_{p1}}. \tag{4.14}$$

The precision of the variance-covariance matrix in PSI can therefore be described by the Dilution of Precision for PSI:

$$\mathrm{DOP_{PSI}} = \sqrt{\det Q_{\varphi}}^{\frac{1}{n}}. \tag{4.15}$$

If the double-difference phase observations would be uncorrelated, $\mathrm{DOP_{PSI}}$ would describe the geometric mean of the standard deviations, see Fig. 4.10.

Similarly, it can be shown that the precision of double-difference displacement estimates can be parameterized by $\mathrm{DOP_{PSI}}$ independent of the spatio-temporal reference. The same type of transformation matrix T is applied to transform to another spatio-temporal difference, and hence the determinant of the variance-covariance matrix of the displacement estimates $Q_{\hat{d}}$ will be invariant as well. An example of the application of DOP_{PSI} as a precision measure can be found in Sect. 7.1.3. Here, it is applied to assess the impact of temporal sampling of a deformation signal, independent of the spatio-temporal reference.

4.4 Measurement Precision

Knowledge about the observation statistics of PSI is essential for the quality description of the deformation parameters of the signal of interest. In Sect. 4.3.1 it has already been noted that the stochastic model of InSAR is not well known a-priori. Hence, this section is dedicated to the validation of InSAR phase observation statistics, which is one of the objectives of the Delft corner reflector experiment (Marinkovic et al., 2004, 2006, 2008; Ketelaar et al., 2004a; van Leijen et al., 2006b). This is a controlled experiment with leveling as an independent validation technique. Since March 2003, five corner reflectors (see Fig. 4.11) have been monitored by leveling and repeat-pass InSAR (ERS-2/Envisat), of which three have survived until 2008. They have all been deployed in a meadow near Delft University of Technology, at 200 meters mutual distance. Besides vali-

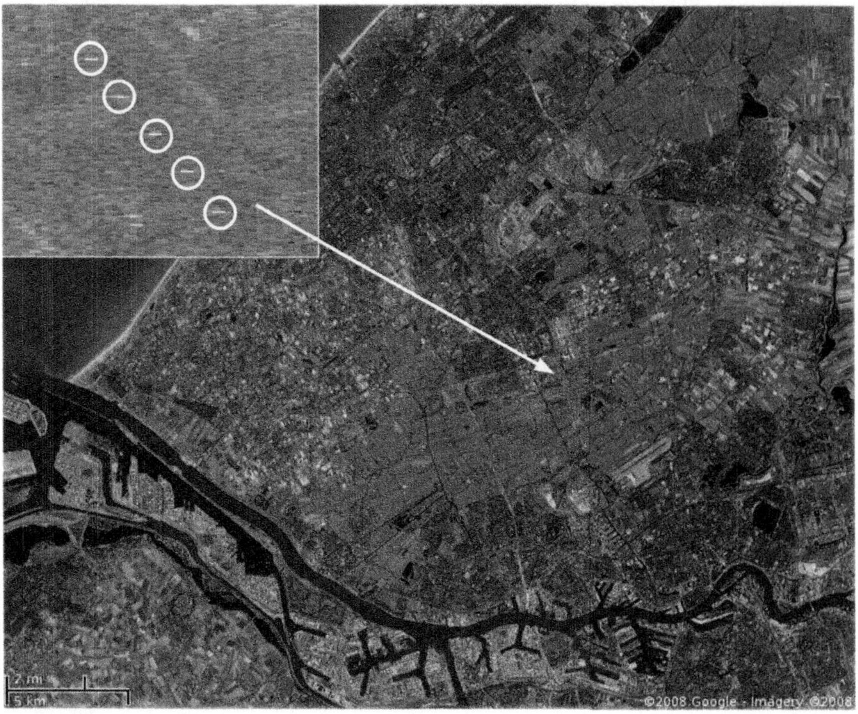

Fig. 4.11 Location and amplitude observations of the five corner reflectors that have been deployed in a meadow near Delft University of Technology (map: © 2008 Google—Imagery © 2008 TerraMetrics)

dation of the measurement precision of PSI, the other objectives are to demonstrate the continuity between ERS-2 and Envisat and to monitor shallow compaction.

This section starts with the set up of the leveling network and addresses the InSAR and leveling double-difference observations. Subsequently, the VCE strategy is explained, which leads to the estimation results for the PSI measurement precision.

4.4.1 Leveling Precision

For each satellite pass, a leveling of the corner reflectors has been performed with a maximum time difference of one day. The heights of the corner reflectors were computed from leveling height difference observations in a redundant network. Using this survey strategy, observational outliers could be detected and removed from the dataset. Two well founded benchmarks were included in this network, see Fig. 4.12: one in a bridge (rp6) and one on the foundation of a highway traffic frame (rp7). Since heights cannot be estimated from height differences, the height of one benchmark has been fixed (rp6). All corner reflector heights have therefore been estimated relative to benchmark rp6.

The measurement precision of leveling height differences is defined as a function of distance. The measurement precision in the corner reflector leveling campaigns is $\sim$1 mm/$\sqrt{\text{km}}$. The precision of the estimated heights is dependent on the measurement precision and the network design. Figure 4.13 depicts the average precision of the corner reflector heights relative to the fixed benchmark rp6. In general, leveling height standard deviations are between 0.5 and 1 mm. The height precision in the first three levelings is slightly worse since no repetitive observations have been taken for each measured rod height. The precision decreases again after July 2006, because the leveling instrument was replaced by a less precise one. Although Fig. 4.13 only depicts the standard deviation of the heights obtained from leveling, the entire variance-covariance matrix is utilized in the validation of the stochastic model of InSAR.

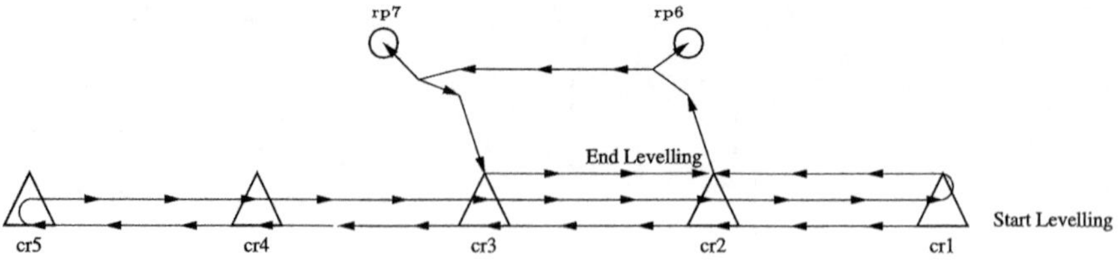

Fig. 4.12 Redundant leveling network for five corner reflectors (triangles). Circles indicate the two reference benchmarks. Arrows indicate the leveling sequence

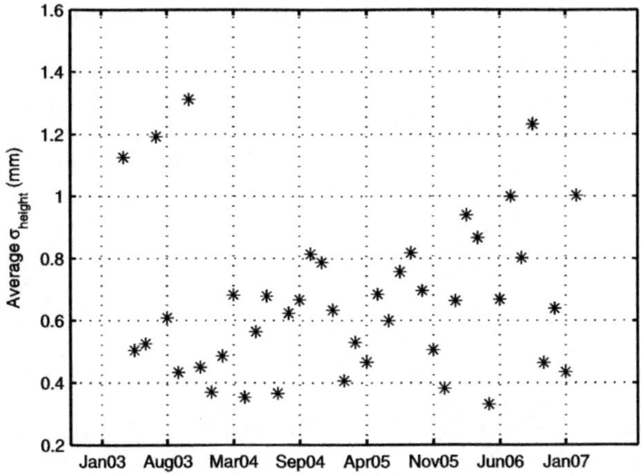

Fig. 4.13 Average standard deviation corner reflector heights in time relative to the fixed benchmark rp6. The measurement precision of the observed height differences is $\sim$1 mm/$\sqrt{\text{km}}$. The precision of the corner reflector heights is $\sim$0.5–1 mm. The height precision in the first three levelings is lower since no repetitive observations have been taken for each measured rod height. The precision decreases again after July 2006, because the leveling instrument has been replaced by a less precise one

4.4.2 InSAR A-priori Measurement Precision

The goal of the corner reflector validation experiment is to validate the stochastic model of InSAR observations. Hence, an a-priori stochastic model is constructed that is validated using VCE (resulting in the a-posteriori stochastic model). The a-priori precision of InSAR observations can be computed as a function of Signal-to-Clutter Ratio (SCR, see (3.1)). The clutter is comprised of both the clutter within the resolution cell of the dominant scatterer and the clutter outside the resolution cell that interferes with the dominant scatterer. Since the SCR is computed from the amplitude observations, amplitude calibration might be required. However, as the location of the corner reflectors and the surrounding clutter can be considered equal, the calibration factors (see Sect. 3.2.2) cancel in the ratio between the signal and the clutter.

SCR estimation has been performed for all five corner reflectors. The corner reflectors behave as point scatterers. Their signature in the spatial domain is a sinc pattern. The signal amplitude value s—the reflection strength of the corner reflector—is the maximum amplitude value of the sinc pattern, see Fig. 4.14. The higher the oversampling factor, the more precise s can be measured.

Since the spatial signature of a corner reflector is a sinc pattern, the interference with sidelobes in range and azimuth direction has to be avoided in the clutter estimation (Hanssen, 2001). Hence, boundaries are set around the signal pattern of the corner reflectors by means of an edge detector. This results in four areas around the corner reflector that are not affected by the point scatterer signal, see Fig. 4.14. The clutter estimation is based on the assumption of ergodicity. As the corner reflec-

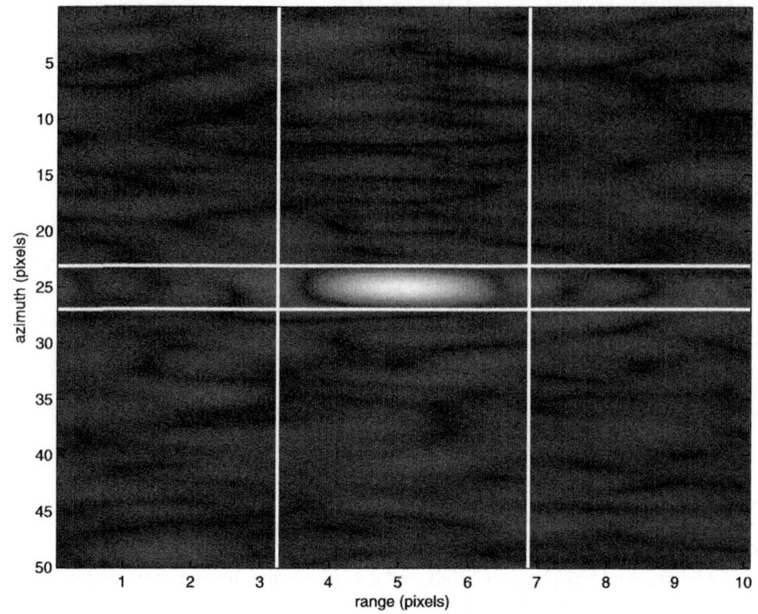

Fig. 4.14 Windows used for SCR estimation. Boundaries are set in range and azimuth direction around the reflection pattern of the corner reflector, to avoid interference of sidelobes in the clutter estimation. Clutter samples are taken in the four quadrants around the corner reflector signal. The oversampling factor is 16

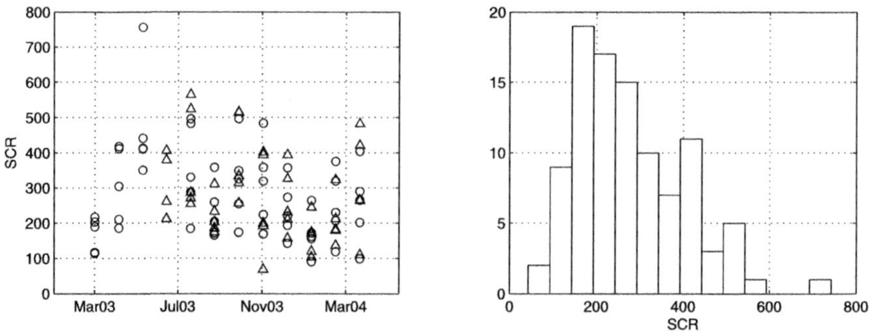

Fig. 4.15 Left: SCR values for corner reflectors 1–5 in time, for ERS-2 (circles) and Envisat (triangles). The SCR values correspond with an observation precision of 0.3–0.6 mm (1-sigma). Right: distribution of the SCR values. The maximum is obtained at a SCR value of ~200

tors have been placed in a large homogeneous field, this assumption is reasonable. To obtain conservative estimates of the phase precision, the mean of the maximum clutter estimates in the four sampling areas is used.

Figure 4.15 shows SCR estimates for the five corner reflectors in time. There is a small seasonal signal visible, which may be caused by clutter changes due to vegetation changes in the surrounding field. The magnitude of the SCR values is of the same order for ERS-2 and Envisat. The majority of the SCR values are between 100 and 500 (20–27 dB), corresponding with SLC phase standard deviations σ_ψ of 0.3–0.6 millimeter (1-sigma). The relation between the displacement estimate D and the double-difference phase observation φ in mm reads: $\varphi = -2D$. Hence, for double-difference line-of-sight displacements this implies a standard deviation of 0.3–0.6 millimeter as well ($\sigma_D = \sqrt{4\sigma_\psi^2/2^2}$).

4.4.3 InSAR and Leveling Double-Difference Displacements

In interferometry, spatial phase differences between reflectors are also differences between two epochs. These double-difference phase observations (in time and space) are the first variates that bear interpretable information. One master has been chosen for both the ERS-2 and Envisat stack (October 2005), and all spatial differences have been referenced to corner reflector 4, which minimizes the spatial distances. As the maximum distance to the reference corner reflector is 200 meters, atmospheric effects are assumed negligible.

Leveling double-differences are orthometric heights along the vertical whereas InSAR double-differences are ellipsoidal heights along the satellite's line of sight, see Fig. 4.16. Assuming that the geoid height does not change in time, the difference between orthometric and ellipsoidal heights cancels in the double-difference, which is further explained in Sect. 7.1.1. Furthermore, it is assumed that the corner reflector displacements only have a component in the vertical direction. Hence, the InSAR double-differences are converted to the vertical as a function of the incidence angle.

The leveling double-differences are a linear combination of the leveling height estimates $\underline{h}$:

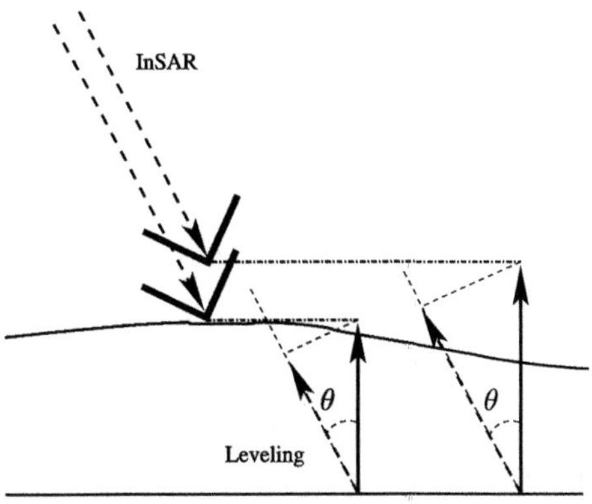

Fig. 4.16 InSAR double-difference observations are in the satellite line of sight; leveling measurements refer to the vertical

$$\underline{d}_{ij}^{t_m t_s} = \begin{bmatrix} 1 & -1 & -1 & 1 \end{bmatrix} \begin{bmatrix} \underline{h}_i^{t_m} \\ \underline{h}_j^{t_m} \\ \underline{h}_i^{t_s} \\ \underline{h}_j^{t_s} \end{bmatrix}, \tag{4.16}$$

where t_m and t_s are master and slave time respectively, and i and j are two leveling points. The double-difference variance is computed by application of the propagation law of variances:

$$\sigma^2_{d_{ij}^{t_m t_s}} = \begin{bmatrix} 1 & -1 & -1 & 1 \end{bmatrix} \begin{bmatrix} \sigma^2_{h_i^{t_m}} & \sigma_{h_i^{t_m} h_j^{t_m}} & 0 & 0 \\ \sigma_{h_j^{t_m} h_i^{t_m}} & \sigma^2_{h_j^{t_m}} & 0 & 0 \\ 0 & 0 & \sigma^2_{h_i^{t_s}} & \sigma_{h_i^{t_s} h_j^{t_s}} \\ 0 & 0 & \sigma_{h_j^{t_s} h_i^{t_s}} & \sigma^2_{h_j^{t_s}} \end{bmatrix} \begin{bmatrix} 1 \\ -1 \\ -1 \\ 1 \end{bmatrix}. \tag{4.17}$$

The leveling height estimates do not correlate between epochs, but correlation will be caused by the spatio-temporal combinations of the double-differences. The structure of the variance-covariance matrix of the leveling heights per epoch depends on the leveling network design.

The InSAR double-difference phase observations are corrected for sub-pixel position (Marinkovic et al., 2006) and topographic height difference and subsequently converted to double-difference displacements in mm along the vertical:

$$\begin{aligned} \underline{d}_{ij}^k = & -\frac{4\pi}{\lambda \cos\theta_i^m}\left(\underline{\varphi}_{ij}^k + \frac{4\pi}{\lambda}\frac{B_i^\perp}{R_i^m \sin\theta_i^m} H_{ij} \right. \\ & \left. -\frac{4\pi}{\lambda}\frac{B_i^\perp}{R_i^m \tan\theta_i^m}\xi_{ij}^0 - \frac{2\pi}{v}(f_{\mathrm{dc},i}^0 - f_{\mathrm{dc},i}^k)\eta_{ij}^0\right), \end{aligned} \tag{4.18}$$

see Sect. 3.4.1 for the definition of the parameters. In this validation experiment, the InSAR double-differences have been unwrapped with respect to the leveling double-differences.

The variance-covariance matrix of InSAR double-differences is a full matrix due to the spatio-temporal combination, see (3.25). If the phase observations in a single SAR acquisition have a variance of σ_ψ^2, the variance-covariance matrix of the double-differences $\underline{\varphi}$ between two corner reflectors has the following structure:

$$Q_\varphi = \sigma_\psi^2 \begin{bmatrix} 4 & 2 & 2 & \dots & 2 \\ 2 & 4 & 2 & \dots & 2 \\ \vdots & & \ddots & & \\ & & & & \\ 2 & 2 & \dots & 2 & 4 \end{bmatrix}. \tag{4.19}$$

Note that correlation between SLC phase observations due to atmospheric signal has been ignored here, due to the short distance between the corner reflectors. The variance-covariance matrices of both the InSAR and leveling double-differences serve as input for the variance component estimation procedure.

4.4.4 Validation of the Stochastic Model

Leveling is a well established measurement technique of which the precision and reliability can be validated in a redundant network set up. For each leveling epoch, erroneous observations have been detected and removed by means of datasnooping. Furthermore, the measurement precision has been estimated from the adjustment residuals, resulting in an update of the initial variance-covariance matrix of the leveling observations Q_h^0 with a variance factor ($Q_h = \sigma^2 Q_h^0$). In the validation of the stochastic model, the leveling variance-covariance matrix is therefore fixed. In this way, the a-priori stochastic model for the InSAR observations can be validated with respect to the independent leveling results.

The a-posteriori precision estimation for InSAR double-differences is performed by means of variance component estimation with the disjunctive group model (Tiberius and Kenselaar, 2003). The observations are partitioned into three uncorrelated groups: leveling, ERS-2, and Envisat double-differences. The variance-covariance matrix of the leveling double-differences remains unchanged, whereas the dispersion of the InSAR double-differences is described by a variance-covariance matrix Q_k with an unknown variance factor σ_k^2. The variance-covariance matrix of the observations reads:

$$Q_y = \begin{bmatrix} Q_{\text{lev}} & 0 & 0 \\ 0 & \sigma_{\text{E2}}^2 Q_{\text{E2}} & 0 \\ 0 & 0 & \sigma_{\text{Ev}}^2 Q_{\text{Ev}} \end{bmatrix}, \tag{4.20}$$

where Q_{lev} is the variance-covariance matrix of the leveling double-differences, E2 refers to the ERS-2 double-differences, and Ev to the Envisat double-differences. Q_{lev}, Q_{E2}, and Q_{Ev} are full matrices, due to the covariances induced by the double-difference combinations.

The functional model that relates the leveling and the InSAR double-difference displacements is set up as a model of condition equations. The conditions that are imposed state that the misclosures between the leveling and InSAR double-difference displacements should be equal to 0:

$$B^T E\{\underline{y}\} = \begin{bmatrix} -I & I & 0 \\ -I & 0 & I \end{bmatrix} \begin{bmatrix} \underline{d}_{ij}^{t_m t_s} \\ \underline{d}_{ij}^{\text{E2}} \\ \underline{d}_{ij}^{\text{Ev}} \end{bmatrix} = 0, \tag{4.21}$$

where B is the condition equation design matrix, i, j are corner reflector indices, t_m, t_s represent master and slave acquisition times, and E2 and Ev indicate the interferometric combinations (ERS-2, Envisat). The condition equations combined with the stochastic model from (4.20) are the input for the variance component estimation.

4.4.5 InSAR A-posteriori Precision

The a-posteriori precision of the InSAR observations, i.e. the variance factors in (4.20), has been estimated from the double-difference displacements of the three corner reflectors that have survived from March 2003 to the end of 2007: corner reflector 3, 4 and 5. These corner reflectors have been deployed in one line, with 200 m distance between the subsequent reflectors. The middle one, corner reflector 4, has been appointed as spatial reference. The time series of double-difference displacements of the corner reflectors 3 and 5 with respect to corner reflector 4 is depicted in Fig. 4.17 (37 ERS-2 and 43 Envisat acquisitions). Clearly, the leveling, ERS-2, and Envisat double-difference displacements are in good agreement. Figure 4.18 shows a scatterplot of the leveling and InSAR double-difference displacements. It is obvious that some outliers are present in the ERS-2 double-difference displacements. It has been chosen to remove these outliers, if the misclosures exceed the standard deviation of double-differences for distributed scatterers, with respect to the displacement time series. This standard deviation follows from the uniform distribution, depending on the effective wavelength:

$$\sigma^2_{d^{ms}_{ij}} = \left(\frac{\lambda}{4\cos(\theta_{\text{inc}})}\right)^2 / 3, \tag{4.22}$$

and is set at 8.8 mm for ERS-2 and Envisat double-differences. This standard deviation is the upper limit for correctly unwrapped double-difference displacements. As

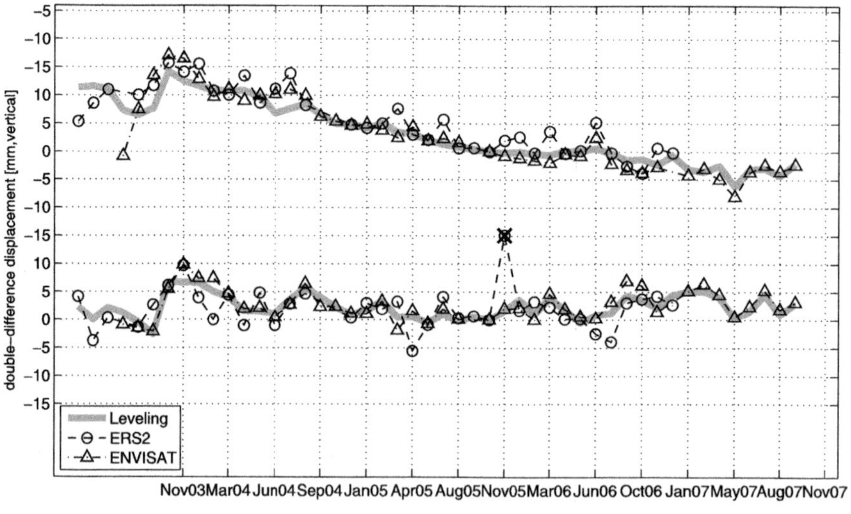

Fig. 4.17 Time series of double-difference displacements for ERS-2 (37 acquisitions), Envisat (43 acquisitions) and leveling (mm, vertical). The time series at the top shows the double-differences between corner reflector 3 and 4; the time series at the bottom shows the double-differences between corner reflector 5 and 4. The outlier in the ERS-2 double-difference displacements has been marked with a cross

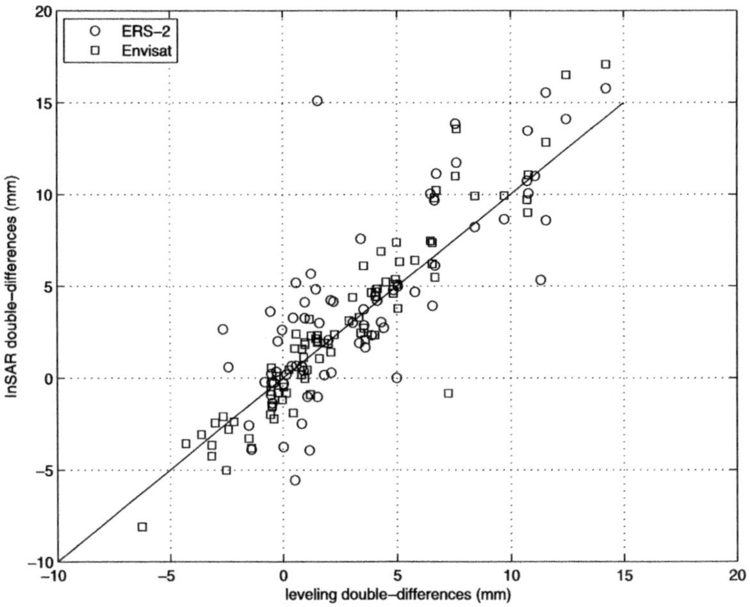

Fig. 4.18 Scatterplot of leveling and ERS-2 (circles) and Envisat (squares) double-difference displacements (37 ERS-2 and 43 Envisat acquisitions). The ERS-2 data suffered from the degraded attitude control of the satellite after 2000. The correlation between leveling and ERS-2 double-difference displacements is 79% before outlier removal and 84% after outlier removal. The correlation between leveling and Envisat double-difference displacements is 94%

a result, one ERS-2 double-difference displacement is rejected and removed from the dataset, which is depicted with a cross in Fig. 4.17. The correlation between leveling and ERS-2 double-difference displacements is 79% before outlier removal and 84% after outlier removal. The correlation between leveling and Envisat double-difference displacements is 94%.

The distribution of the misclosures between leveling and InSAR is shown in Fig. 4.19. The a-posteriori precision of InSAR double-differences is obtained from VCE as described in Sect. 4.4.4. It has resulted in an average double-difference displacement precision of 3.0 and 1.6 mm for ERS-2 and Envisat respectively, see Table 4.2.

Although it can be concluded that the a-priori InSAR precision based on SCR (see Fig. 4.15) is overestimated, the standard deviation of InSAR double-difference displacements is in the order of several millimeters. The precision of Envisat double-difference displacements is equal to the leveling precision, which shows that InSAR can be applied for deformation monitoring. This is strengthened by the correlation coefficient of 0.94 between the leveling and Envisat double-difference displacements. A likely cause for the lower precision of ERS-2 are the large Doppler deviations in the time series (from −1700 to +3600 Hz). Hence, the phase observations are very sensitive to the sub-pixel position. The influence of the sub-pixel position has been investigated by Perissin (2006) and Marinkovic et al. (2006). An-

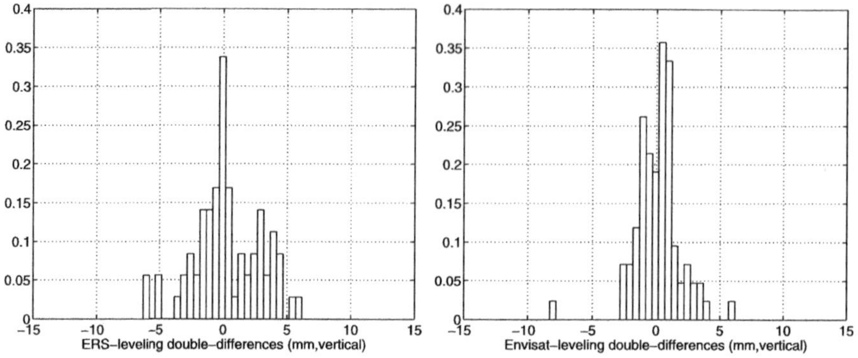

Fig. 4.19 Double-difference misclosures between ERS-2 (left), and leveling and Envisat (right) and leveling in mm converted to the vertical. The a-posteriori double-difference displacement precision is 3.0 and 1.6 mm for ERS-2 and Envisat respectively

Table 4.2 Average standard deviation of double-difference displacements after VCE for ERS-2 and Envisat. The standard deviation of the estimates obtained from VCE is 0.2 mm

	leveling	ERS-2	Envisat
σ_d (mm)	1.5	3.0	1.6
σ_{σ_d} VCE (mm)	–	0.2	0.2

other validation experiment that uses dihedral reflectors and GPS measurements is described by Ferretti et al. (2007), which claims sub-millimeter precision of InSAR displacement estimates, both in vertical and horizontal direction.

4.5 Idealization Precision for Deformation Monitoring

In the introduction of this chapter, it has been explained that quality control for deformation monitoring consists of two components: precision and reliability of the measurement technique, and the relation of the deformation estimates to the signal of interest.

The precision and reliability of PSI as a measurement technique has been addressed in the Sects. 4.1 to 4.4. The influence of model errors on the PS height and velocity estimates has been investigated. Velocity errors in the order of 0.5 mm/year can be expected if the azimuth sub-pixel position is not estimated. Orbit errors can result in velocity errors of ~1 mm/year over a full SLC extent. Although these results are based on the optimistic scenario of correct phase unwrapping, and the reliability of the deformation estimates needs to be further investigated (see Chap. 5), it can be stated that the precision of PSI as a measurement technique is at mm level. In a controlled corner reflector experiment, it has been shown that the double-difference displacement precision for Envisat is 1.6 mm, similar to the precision of leveling double-difference displacements (1.5 mm).

The availability of precise geodetic measurements from techniques such as PSI and leveling is not necessarily sufficient for a precise and reliable estimation of the deformation signal of interest. Hence, this section focuses on the relation between the deformation estimates and the signal of interest. It gives an overview of the available tools in PSI to enhance the estimation of the deformation signal of interest. In this context, the concept of *idealization precision* is used as a measure for the match between the measurements and the displacements that are caused by the signal of interest.

In classical geodesy, the idealization precision gives an indication of the identification precision of a point in the terrain. Points with a high idealization precision, such as the corner of a house, can be sharply identified. Compared to the corner of a house, the middle of a canal has a lower idealization precision. The same concept is applicable to deformation parameter estimation from geodetic measurements. For example: if the signal of interest, subsidence due to gas extraction, is contaminated by shallow subsurface deformation of a similar magnitude, it has a low idealization precision.

Although idealization precision in deformation estimation plays a role in each geodetic measurement technique, it is more prominent in PSI than in traditional techniques such as leveling or GPS, due to the physical properties of the measurement points. Leveling uses well defined benchmarks that are for example established in a building founded on a stable subsurface layer. In case of SAR reflections, it is complicated to identify the physical measurement target(s) and the type of the reflection due to the limited precision of height estimates and geocoding (in the order of meters, see Perissin (2006)). Furthermore, the earth surface can be affected by spatial and temporal changes that originate from multiple deformation causes. Since earth surface deformation is observed regardless of the deformation mechanism by a remote sensing technique such as InSAR, it may be a challenge to discriminate the deformation signal of interest.

After the explanation of deformation causes in Sect. 4.5.1, Sect. 4.5.2 will address the tools that are available in PSI to increase knowledge on the physical PS properties and the reflection type. Subsequently, Sect. 4.5.3 will invesigate idealization precision in the deformation parameter space. Here, knowledge on the spatio-temporal correlation of the deformation signal of interest is utilized to perform a PS selection according to the deformation regime.

4.5.1 Deformation Regimes

Since radar satellites monitor movements of scatterers from space, the interferometric phase differences may represent deformation caused by a variety of deformation mechanisms. The signal of interest, subsidence due to gas extraction, may be contaminated by other spatio-temporal displacements. Thus, the deformation contribution

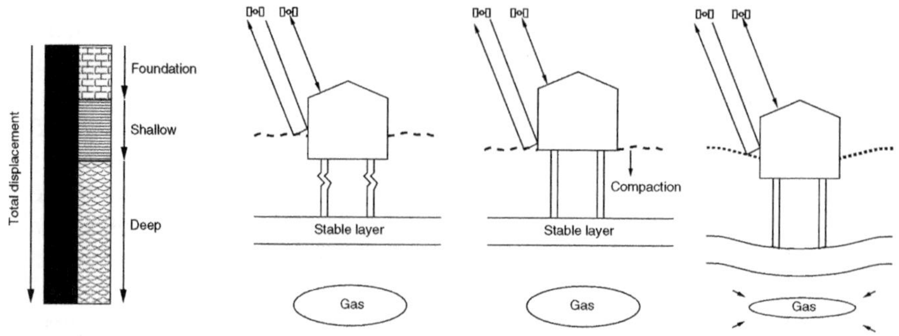

Fig. 4.20 Deformation regimes and their effect on single and double-bounce reflections: (left) structural instabilities (foundation), affecting only the single-bounce reflection; (middle) shallow mass displacement (compaction), affecting only the double-bounce reflection; (right) deep mass displacement (gas extraction), affecting both the single-bounce and the double-bounce reflection

in a PS phase observation can be a superposition of displacements due to several *deformation regimes*, such as:

- structural instabilities (including foundation),
- shallow mass displacements (ground water level variation and compaction),
- deep mass displacements (gas, oil and mineral extraction).

All three deformation regimes are present in the Netherlands, and more specific in the subsiding area above the Groningen gas field, see Sect. 6.5.

To be able to relate the movement of a PS to its driving mechanisms it is important to consider the possible scattering characteristics. A PS can be a dominantly specular (single-bounce) reflection from a building, but it might as well be a dihedral reflection (double-bounce, curb-to-wall), comparable to a dihedral corner reflector. This can be extended to multiple bounce effects, for example triple-bounce, comparable to a trihedral corner reflector. Figure 4.20 shows possible deformation causes and the difference between a single-bounce and a double-bounce reflection. If the physical nature of the PS is unknown, soil compaction can easily be incorrectly interpreted as gas extraction.

4.5.2 PS Characterization

This section addresses the techniques in PSI to enhance the identification of the physical PS properties and the reflection type. A PS displacement may be caused by (a superposition of) several deformation regimes. For the separation of deformation regimes, knowledge about the reflection type is required. If the signal of interest is subsidence due to deep subsurface displacements, it is essential to know if the PS displacement represents a direct reflection from a well founded building on a deep subsurface layer, or a double-bounce reflection with the surroundings affected by shallow compaction.

Fig. 4.21 A possible double-bounce reflection (PS velocity −9 mm/year). Since the pavement around the building is subsiding, the SAR reflection is likely to be a curb-to-wall reflection. The orientation of the building towards the satellite look direction confirms that a double-bounce reflection is possible

The precision of the horizontal location of a PS is not sufficient to determine whether a PS stems from ground level or roof. PS 3D positioning accuracy in the order of 1 meter can be obtained, see Perissin (2006). If the PS radar coordinates are only provided at pixel level, the accuracy depends on the ground resolution and the precision of the PS height estimates. For an oversampling factor of 2, the ground resolution of a pixel is 2×10 meters in azimuth and range direction (ERS, Envisat). The precision of the height estimates determines the precision of the geolocation. The relation between height H and horizontal position x can be approximated by:

$$\delta x = \frac{\delta H}{\tan(\theta_i)}, \tag{4.23}$$

where θ_i is the incidence angle. The factor between the deviation in the height estimate and the horizontal location is ~ 2.5. Hence, the horizontal location can be easily more than 10 meters off, if the sub-pixel position is not taken into account.

For better PS identification, several techniques exists in PSI. The first one has already been mentioned: the PS height estimates. Although their precision may be limited to 1 meter, it can be sufficient to determine whether the PS reflection stems from ground level or from a roof top. The other two techniques use respectively

Fig. 4.22 PS targets in the Netherlands: roofs that are aligned perpendicular to the satellite look direction. These targets are most likely represented by a direct specular reflection

Fig. 4.23 More PS targets in the Netherlands: (left) dike covered with basalt blocks that is a distributed PS, (right) sluice that is most likely represented by a multi-bounce reflection

the PS reflectivity pattern as a function of viewing geometry, and polarimetric Envisat observations. The common goal of these methods is to distinguish between direct specular reflections and multiple bounce (mainly dihedral) reflections. The application of PS characterization in this research is based on the assumption that specular reflections from settled buildings founded on a deep subsurface layer are the most suitable targets for the estimation of subsidence due to deep subsurface displacements, see Sect. 6.5.

4.5.2.1 PS Heights

PS heights can be used to estimate whether a reflection stems from the top of a building or ground level. The PS heights are incorporated as unknown parameters in the functional model of PSI, see (3.11). Their precision depends on the precision of the double-difference phase observations, number of acquisitions (redundancy), and the acquisition geometry (distribution of perpendicular baselines). Perissin (2006) has shown that a height precision of several decimeters can be achieved, provided that the sub-pixel location of the target is known.

The accuracy of the sub-pixel location, especially the range coordinate, has nevertheless a significant impact on the height estimates, see Sect. 4.2.1. The phase contribution due to the PS height reads:

$$\underline{\varphi}^k_{H,ij} = -\frac{4\pi}{\lambda} \frac{B_i^{\perp}}{R_i^m \sin\theta_i^m} H_{ij}, \tag{4.24}$$

whereas the phase contribution due to the slant-range sub-pixel position reads:

$$\underline{\varphi}^k_{\eta,ij} = +\frac{4\pi}{\lambda} \frac{B_i^{\perp}}{R_i^m \tan\theta_i^m} \eta_{ij}^m. \tag{4.25}$$

Since there is a direct linear relation between the PS height and the range sub-pixel position, the deviation in PS height due to an incorrect range coordinate can be expressed as:

$$\delta H = -\cos\theta_i \cdot \delta\eta. \tag{4.26}$$

For a slant-range pixel spacing of 7.9 meters (ERS), this implies a PS height deviation of −7.3 meter. If a stack of SAR acquisitions is oversampled by a factor of 2 and no additional range sub-pixel position estimation is performed, height deviations of ~3.5 m can be expected, which is confirmed by Fig. 4.1 in Sect. 4.2.1. For a decrease of the height deviations, the sub-pixel position of the PS targets needs to be estimated more accurately. However, this is not straightforward. Because of the linear dependency between the PS height and the range sub-pixel position, the range sub-pixel position cannot be incorporated as an additional unknown parameter in the functional model. This leaves only the option of estimation of the range sub-pixel position based on the (oversampled) amplitude observations, like it has been performed for the corner reflectors, see Sect. 4.4 and Fig. 4.14.

It is essential that sidelobes are removed from the PSI results, prior to PS characterization based on PS height estimates. The phase observations of sidelobes are duplicates of the phase observations of the actual target, but they have been assigned to the wrong range bin, and hence their height estimates are incorrect, see Sect. 4.3. Figure 4.24 shows an example of height variations of sidelobes: they are in the range of −20 to +20 meters.

Sidelobe removal procedures can be subdivided into two groups:

1. sidelobe removal based on amplitude observations (convolution with a sinc pattern), and
2. removal of dependent pixels based on the differential phase observations (Perissin, 2006).

Ideal point targets appear as a sinc pattern in space. Therefore they can be detected by convolution with a sinc kernel. The full time series of acquisitions can be utilized by complex multiplication of oversampled PS patches for the entire stack. In this way, clutter is suppressed and point targets become more prominent. A refined

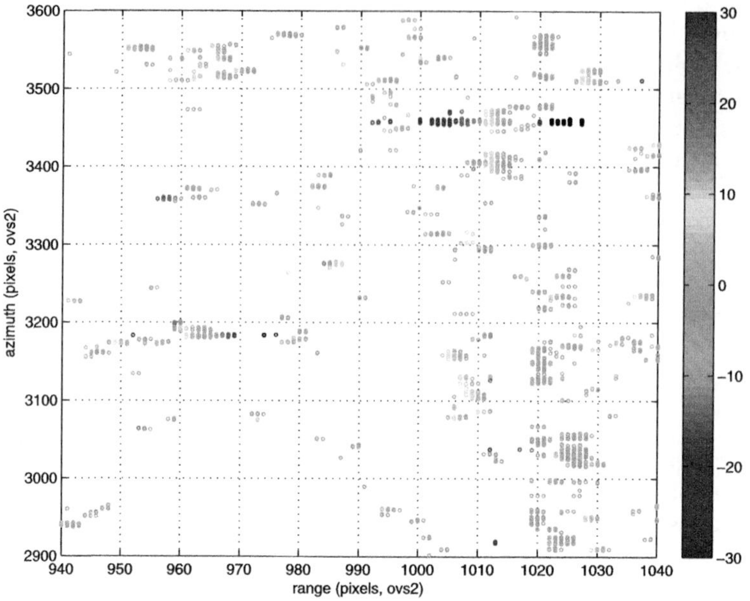

Fig. 4.24 PS height estimations including sidelobe observations. The height deviations of the sidelobes are dependent on their range location

estimation of the sub-pixel location of the point target can be determined by additional oversampling. This procedure has been applied for the study described in Ketelaar et al. (2005).

Perissin (2006) has developed a method to detect and remove dependent pixels that is based on the reversal of the phase observations between subsequent sidelobes. It is based on a correlation measure of phase observations between pixels in range and azimuth direction. By using a stack of acquisitions, the detection accuracy of dependent pixels can be increased.

Sidelobe removal based on amplitude observations has the disadvantage that the spatial pattern of natural targets deviates from the sinc pattern of an ideal point target. Hence, sidelobes with strong reflections can still be detected as PS targets, dependent on the threshold for the similarity between the amplitude pattern and the sinc pattern. Sidelobe removal based on phase correlation of dependent pixels is influenced by the phase correlation threshold, see Fig. 4.25. Perissin (2006) shows however that the likelihood of erroneously rejecting an independent pixel is extremely low for PS targets with a correlation threshold of 0.8. Hence, sidelobe removal based on phase observations is preferred. Optionally, an initial selection is performed by choosing pixels that correspond with local amplitude maxima.

Summarizing: PS heights can be used for PS characterization after removal of sidelobes. Besides that, the precision of the height estimates needs to be quantified as a function of the precision of the phase observations, the number of acquisitions, the acquisition geometry, and the accuracy of the range sub-pixel position.

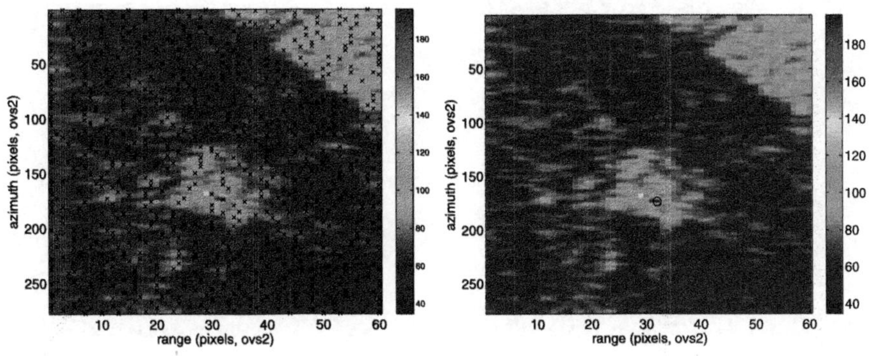

Fig. 4.25 Sidelobe removal based on the differential phase observations. The number of independent pixels is dependent on the correlation threshold. Left: a correlation threshold of 0.2. Right: a correlation threshold of 0.8. The higher the threshold, the less targets are incorrectly identified as independent targets

4.5.2.2 PS Reflectivity as a Function of Viewing Geometry

Another method for PS characterization is based on reflectivity as a function of acquisition geometry (Ferretti et al., 2005). This method relies on the assumption that specular reflections are observed from a smaller range of viewing angles than dihedral reflections. It utilizes the variations in incidence angle (perpendicular baseline) and squint angle (Doppler centroid frequency) to identify the PS reflectivity behavior.

Ferretti et al. (2005) model the PS amplitude observations as a sinc that is a function of perpendicular baseline and Doppler centroid frequency. The unknown parameters are (i) the target extension, and (ii) the perpendicular baseline and Doppler centroid frequency of maximum reflection (the position of the main lobe). The target extension is the target size across-track and along-track in meters, and is inversely proportional to the sinc width.

Figure 4.27 shows the reflectivity of a target in the northern part of the Netherlands as a function of incidence and squint angle. The observed amplitudes have first been converted to backscatter coefficients following the calibration procedure of Laur et al. (2002), but omitting the corrections for a different viewing geometry. These backscatter coefficients can be considered as normalized intensity (reflectivity) observations, that are the input for the estimation procedure to obtain the best fit for the reflectivity pattern. The best fit of the reflectivity pattern is detected in a predefined search space of the unknown parameters. The target depicted in Fig. 4.27 is most likely a dihedral, since its reflection stays strong over the entire range of incidence angles. The position of the main lobe as a function of squint angle deviates from 0, which indicates that the orientation of the target differs from the satellite ground track of the master.

Fig. 4.26 Left: the location of the scatterer from Fig. 4.27 in the multi-image reflectivity map. The SAR data has been provided by ESA. Right: the corresponding geographic location (map: © 2008 Google—Imagery © 2008 DigitalGlobe, GeoContent, Aerodata International Surveys, GeoEye)

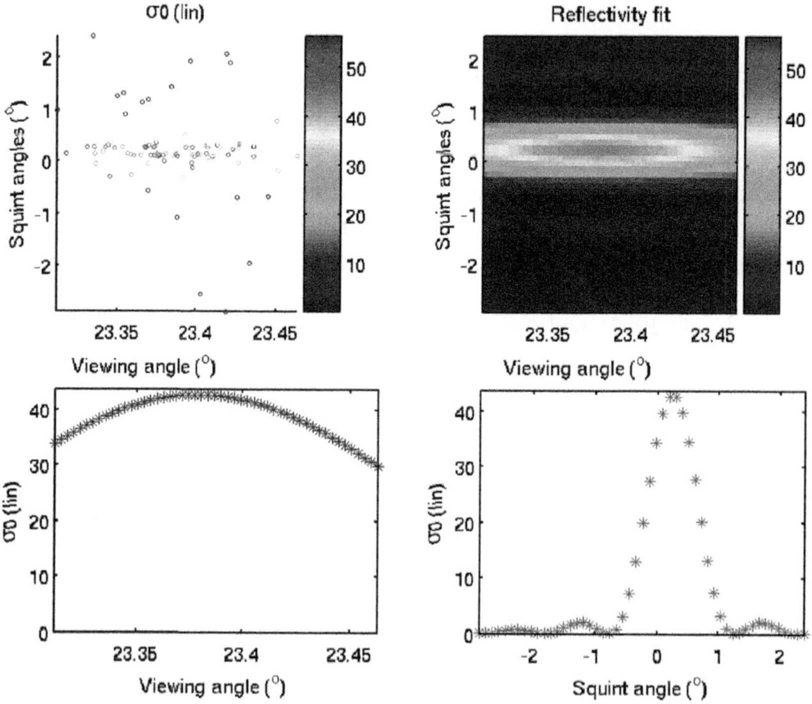

Fig. 4.27 Reflectivity pattern fit: normalized intensity observations as a function of incidence and squint angle, reflectivity pattern fit and its profiles in range (incidence angle) and azimuth (squint angle) direction

4.5.2.3 Alternating Polarization

Polarimetry can be used to distinguish between even and odd bounce scatterers (van Zyl, 1989; Hoekman and Quinones, 1998). In case of an odd number of bounces, the phase angle between HH and VV polarization is 180°, whereas for an even number of bounces it equals to 0°. Hence, polarimetric data can enhance the classification of PS into specular (or trihedral) and dihedral scatterers.

Contrary to ERS-1 and ERS-2, Envisat is able to acquire dual polarization data. The SAR data is acquired in bursts of alternating polarization (HH/VV, HH/HV or VV/VH), which is referred to as the Alternating Polarization (AP) mode. In this way, two scenes that cover the same area are acquired simultaneously in different polarization modes. Inglada et al. (2004) explain how the azimuth spectrum is subdivided between the two polarimetric modes. It is also shown that the HH and VV spectra are shifted by a quarter of the bandwidth, and hence the phase difference maxima will be located around 0.5π and 1.5π, in stead of 0 and π.

The information in an Envisat AP acquisition (HH/VV) is depicted in Fig. 4.28. From the coregistered HH and VV scenes, the auto-interferogram can be computed. From this auto-interferogram the HH/VV phase difference can be obtained for PS candidates. Figure 4.28 shows an increase of distinctive behavior of HH-VV phase differences for pixels with higher amplitudes (potential PS candidates) in the auto-interferogram.

The target information obtained from the Envisat AP acquisitions can be coregistered with the ERS PSI results. Although PS are less sensitive for changes in

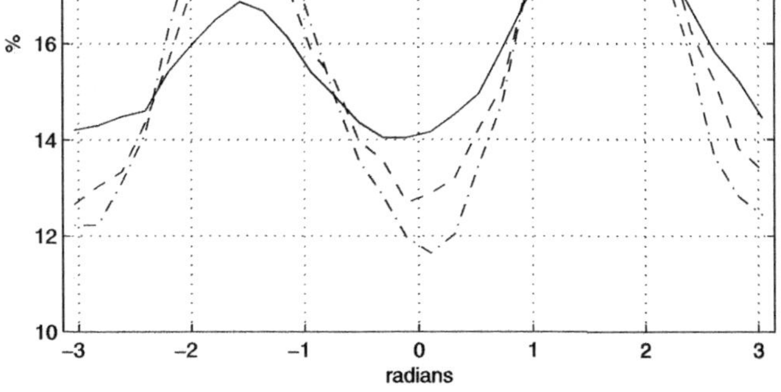

Fig. 4.28 HH-VV phase difference histograms for pixel selection based on amplitude (square root of the amplitudes in the auto-interferogram). The higher the amplitude, the more distinctive the histogram peaks at 0.5π and 1.5π. This demonstrates the information on reflection type that is contained by polarimetric data

viewing angle, the information in the AP data is optimally used if the number of PS targets that are viewed by both ERS and Envisat is optimized. The number of targets that are observed by both sensors is dependent on the perpendicular baseline of the ERS–Envisat combination. The optimal ERS–Envisat baseline compensates for the frequency difference Δf of 31 MHz between the two sensors (Perissin, 2006):

$$B_{\perp} = \frac{\Delta f}{f_0} R_0 \tan(\theta - \alpha), \tag{4.27}$$

where $\Delta f = f_{\text{Envisat}} - f_{\text{ERS}}$. The baseline that results in the same viewing angle is $\sim$2 kilometers, where the ERS image is considered as the master image. The orbital track of Envisat is located to the left of ERS.

4.5.3 *The Use of A-priori Knowledge on the Deformation Signal*

Besides increasing the knowledge on the physical PS properties, the idealization precision for deformation monitoring can also be improved by identification of the observed deformation regime(s). Subsequently, the functional or stochastic model can be adapted to incorporate multiple deformation regimes. If there is a specific signal of interest, such as subsidence due to gas extraction, selection of a subset of PS that is likely to represent that deformation regime, may be sufficient. Another option is to decompose the PS displacements into components due to different deformation regimes. This section will address these strategies and comment on their applicability.

4.5.3.1 PS Selection Based on Spatio-temporal Behavior of the Signal of Interest

Depending on the signal of interest, PS can be selected based on spatial correlation with their neighboring targets. In fact, such a selection is often already performed when selecting PS based on the residual phase coherence (Ferretti et al., 2001) in the PSI estimation. The PS that have a phase behavior in time according to the modeled deformation (for example a linear displacement), are likely to have a higher coherence than a PS that deviates from the modeled deformation. This section starts however from the PSI results in which all deformation regimes are still present. PS will subsequently be selected based on the a-priori knowledge of the spatio-temporal behavior of the signal of interest. Two methods are addressed:

1. outlier removal in grid cells where the displacements are assumed constant,
2. Kriging cross-validation using the covariance function of the deformation signal of interest.

The first method utilizes a quadtree decomposition of (a prognosis of) the signal of interest, see Fig. 4.29. Within each quadtree cell, the deformation estimates are

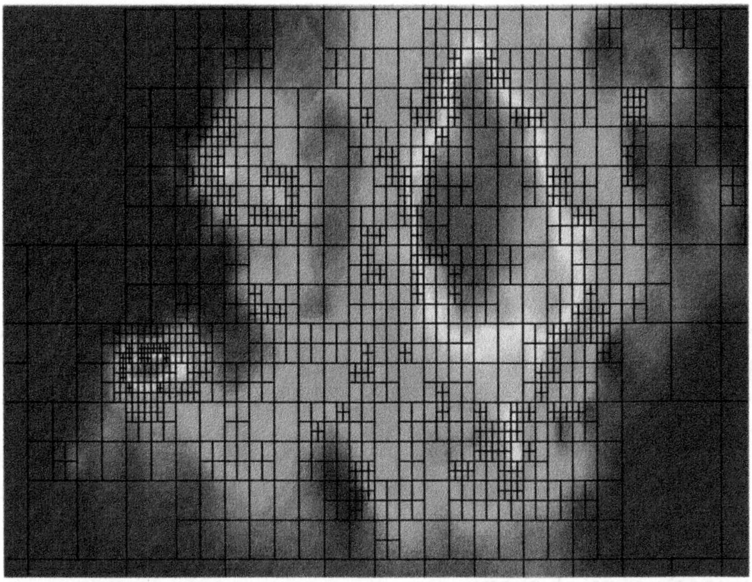

Fig. 4.29 Quadtree decomposition of interpolated PS velocities. At the slopes of the subsidence signal, the quadtree grid cells are smaller to capture the characteristics of the subsidence pattern.

considered constant. A datasnooping procedure is applied to remove outliers, and PS that are spatially correlated remain. The teststatistic for datasnooping is the $\underline{w}$-teststatistic (Teunissen, 2000b):

$$\underline{w} = \frac{c_y^T Q_y^{-1} \underline{\hat{e}}}{\sqrt{c_y^T Q_y^{-1} Q_{\hat{e}} Q_y^{-1} c_y}}, \tag{4.28}$$

where Q_y is the variance-covariance matrix of the observations, and $\underline{\hat{e}}$ is the vector of least-squares residuals with the corresponding variance-covariance matrix $Q_{\hat{e}}$. The vector c_y selects one observation for which the $\underline{w}$-teststatistic is computed, see Sect. 2.3.1. If the observations are uncorrelated, then the $\underline{w}$-teststatistic for the ith observation simplifies to:

$$\underline{w}_i = \frac{\underline{\hat{e}}_i}{\sigma_{\hat{e}_i}} = \frac{\underline{y}_i - \underline{\hat{y}}_i}{\sigma_{\hat{e}_i}}, \tag{4.29}$$

where $\underline{\hat{y}}_i$ is the least-squares estimate of the ith observation.

The second method uses the Kriging cross-validation teststatistic (Wackernagel, 1998):

$$Z^*_{OK}(x_0) = \sum_{k=1}^{n} w_k Z(x_k), \qquad T_i = \frac{Z(x_i) - Z^*_{OK}(x_i)}{\sqrt{\sigma_i^2 + \sigma_y^2}}, \tag{4.30}$$

where:

$Z^*_{OK}(x_0)$	Ordinary Kriging value on location x_0, based on n neighboring points,
w_k	Kriging weight for a neighboring point k based on the covariance function,
$Z(x_i)$	observation for point i,
$Z^*_{OK}(x_i)$	Kriging estimate for point i, based on n neighboring points excluding i,
T_i	cross-validation teststatistic for point i,
σ_i^2	Kriging variance,
σ_y^2	measurement variance.

For each PS, a least-squares interpolated (Ordinary Kriging) displacement or velocity is calculated, based on the surrounding PS estimates, using the covariance function of the deformation signal to determine the weights. The discrepancy of this interpolated displacement or velocity with the actual estimate divided by the measurement and Kriging standard deviation forms the cross-validation teststatistic, which has a standard normal distribution. The Ordinary Kriging estimator is unbiased, as the condition is applied that the sum of the Kriging weights should be equal to 1. Only PS with a cross-validation teststatistic that is lower than a certain threshold are selected: their deformation behavior is in agreement with neighboring targets and the signal of interest.

In fact, datasnooping in quadtree grid cells and Ordinary Kriging are similar. Ordinary Kriging is based on a constant, but unknown mean, similar to the assumption of constant deformation within the quadtree grid cells. Moreover, the covariance function that describes the spatio-temporal behavior of the signal of interest can be easily incorporated in the variance-covariance matrix of the deformation estimates that are the input for datasnooping. The system of equations for the estimation of the deformation signal then can be formulated as:

$$\underline{y} = Ax + \underline{e} = Ax + \underline{s} + \underline{n}; \qquad Q_y = Q_{ss} + Q_{nn}, \tag{4.31}$$

where the design matrix A describes the functional relation between the observations $\underline{y}$ and the unknown deformation parameters x. The part of the deformation signal that is stochastically modeled is denoted by $\underline{s}$. Measurement noise is represented by $\underline{n}$. Besides the relation between Kriging cross-validation teststatistics and $\underline{w}$-teststatistics, Kriging methodologies are related to the $\underline{y}^R$ variates theory (Teunissen, 2000a) and Best Linear Unbiased Prediction (BLUP) (Teunissen, 2007). The Best Linear Unbiased Predictor of the stochastically modeled deformation in (4.31) reads:

$$\hat{\underline{s}} = Q_{ss} Q_y^{-1} (\underline{y} - A\hat{\underline{x}}), \tag{4.32}$$

which is a weighted linear function of the observables. The BLUP is unbiased and has minimum variance.

In stead of using a model-driven approach (spatio-temporal behavior of the deformation regime) different deformation patterns can also be distinguished in a data-driven way. The data-driven method clusters PS that have similar displacement time series (Ketelaar and Hanssen, 2003). The deformation regimes that drive the displacements for each cluster have to be identified afterwards. Although this method may not be efficient for the estimation of a single deformation signal of interest, it has the advantage that is does not utilize any a-priori assumptions on the underlying deformation models.

4.5.3.2 Decomposition of PS Displacements According to Deformation Regime

In the selection of PS based on the spatio-temporal correlation of the signal of interest, one rejects PS that do contain useful information, but are likely to be contaminated by other deformation regimes. If it would be possible to estimate deformation regime *components*, one would optimally benefit from all PSI displacement estimates. However, such a procedure requires knowledge about the functional model of all superposed deformation regimes, which behavior can be very complex. The number of parameters needed to characterize them in the functional model may exceed the number of observations and lead to an unsolvable system. Therefore, deformation regimes are often modeled stochastically based on their spatio-temporal behavior.

Modeling deformation regimes stochastically using covariance functions to construct a variance-covariance matrix for the observations is strongly related to techniques such as factorial Kriging and Principal Component Analysis (PCA) (van Meirvenne and Goovaerts, 2002). In case of factorial Kriging, multiple variograms are constructed to map each spatial component separately. The total variogram is a superposition of variograms with different correlation lengths. To use factorial Kriging, knowledge of the different correlation lengths of the deformation regimes is required. In PCA, the eigenvalues and eigenvectors of the variance-covariance matrix specify the magnitude and direction of the stochastic processes. A disadvantage of PCA is that physical interpretation of the eigenvalues and eigenvectors is often not straightforward. Other methodologies for the determination of the contribution of different deformation regimes are described by Langbein and Johnson (1997), where the presence of time-correlated noise besides white noise in geodetic time series is distinguished. Spectral estimation of the noise components in a signal is investigated here. Since spectral estimation has the disadvantage that regular sampling is required, a maximum likelihood technique that optimizes the magnitudes of the noise types is investigated (ibid.).

To investigate if the limitations of the above mentioned methods can be overcome, the applicability of variance component estimation (VCE) (Teunissen, 1988) is considered in this section. The starting point is the variance-covariance matrix in the PSI estimation, see (3.24):

$$Q_y = W(Q_n + Q_{\text{defo}} + Q_{\text{atmo}})W^T, \tag{4.33}$$

and its Taylor series decomposition with respect to the stochastic parameters in Sect. 4.3.3. In the presence of multiple deformation regimes, Q_{defo} is further subdivided:

$$Q_{\text{defo}} = \sum_{d=1}^{D} Q_{s_d}, \tag{4.34}$$

where D is the number of deformation regimes. As is pointed out in Sect. 4.3.3, the stochastic parameters of the covariance functions of the different deformation regimes should be independently estimable. This is a major limitation, since it implies that deformation regimes with a similar spatio-temporal behavior, but with different correlation lengths, cannot be estimated independently. Another requirement is imposed on the spatial and temporal sampling of a deformation regime. Correlation lengths smaller than twice the distance between PS cannot be estimated.

To verify the estimability of the variance factors, simulations have been carried out in the same way as described in Sect. 4.3.3 but with additional focus on spatio-temporal sampling and multiple deformation regimes. The VCE simulations focus on the precision of the variance component estimates. The ratio between the precision and the estimated value of a variance factor determines its level of significance and subsequently the separability of the deformation regime it describes. The varying elements in the simulations are:

- spatial and temporal sampling frequency (number of measurement points and epochs),
- measurement noise and number of deformation regime parameters.

As most of the cofactor matrices of the deformation regimes are full matrices, VCE requires a lot of processing time. The simulations have therefore been restricted to small networks.

If the number of variance components to be estimated increases while the network redundancy stays the same, the precision of the estimated variance components decreases, as can be deduced from Fig. 4.30. The higher the spatial and temporal sampling frequency, the higher the redundancy, and the higher the level of significance of the variance factor estimates.

The variance components can only be estimated if they are independent. If two deformation regimes with the same covariance function are incorporated in VCE,

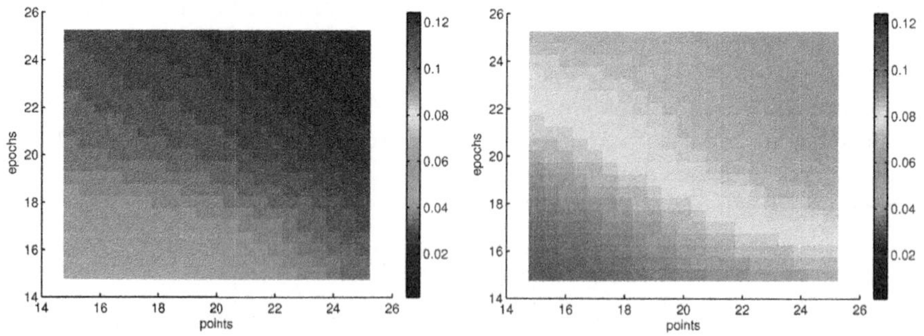

Fig. 4.30 Precision of a stochastic model parameter in the estimation of four (left) stochastic model parameters (measurement noise and spatio-temporal model imperfections) and six (right) stochastic model parameters (measurement noise, autonomous temporal movements, and spatio-temporal model imperfections). The precision is varying with the number of measurement points and the number of epochs (redundancy)

the matrix N in (2.12) will not be of full rank, and therefore the solution of the stochastic parameters can not be uniquely obtained.

Hence, to conclude, the use of a search space of correlation lengths for deformation regimes with a similar covariance function is investigated. The deformation signal is in this situation considered as a superposition of signals of fixed correlation lengths L with unknown magnitudes σ^2:

$$Q_{\text{defo}} = \sum_{d=1}^{D} (\sigma_d^2 Q_{L_d}). \tag{4.35}$$

By means of VCE, the magnitudes of the signals of different correlation lengths can be estimated. If a signal is not present, its magnitude should be 0. Figure 4.31 shows a simulation of superposed signals with correlation lengths of 0, 3 and 10 kilometer and variances 4^2, 3^2, and 7^2 mm^2 respectively. Figure 4.31 shows the estimated variance factors of the three regimes if these correlation lengths would have been known a-priori. They match approximately the input variances of the superposed signals. When the correlation lengths are not known and VCE is performed in a search space of correlation lengths (0, 3, 6, 9, 12 km), it can be seen that the three regimes are less prominent. The search space regimes are correlating, and moreover since no positiveness constraints are imposed on the variance factors, the original regimes cannot be unambiguously distinguished.

This clearly indicates the limitation of the separation of displacement components using VCE. Components due to different deformation regimes can be separated, provided that their stochastic model parameters are independent. An example will be shown in Sect. 6.5.4.2. The separation of deformation regimes with a similar covariance function however relies on the geophysical interpretation of the total sum of all spatially correlated deformation signals.

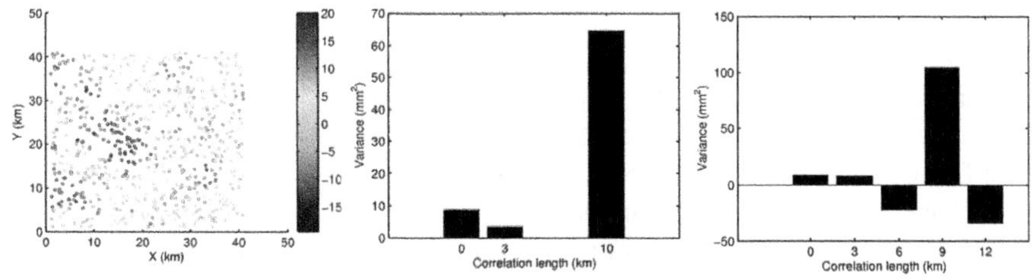

Fig. 4.31 Variance factor (magnitude) estimation of superposed deformation regimes with spatial correlation lengths of 0, 3, and 10 kilometer (left). A search space of spatial correlation lengths is defined. If the correlation lengths in the search space are restricted to the correlation lengths that are present in the signal, the magnitudes of the deformation regimes are correctly estimated (middle). This does not hold for an arbitrary search space of correlation lengths (right). Hence, the unknown correlation lengths of the deformation regimes cannot be estimated by an arbitrary pre-defined search space of correlation lengths

4.6 Conclusions

This chapter has shown the achievable precision and reliability of PSI as a measurement technique. Furthermore, the concept of idealization precision for the estimation of the deformation signal of interest has been introduced. Subsequently, the applicability of available techniques to increase the idealization precision for deformation monitoring within PSI has been described.

Since the accuracy of a measurement technique is not only determined by its precision, the effect of model errors on the PSI parameter estimates has been investigated. Shortcomings in the functional model have been evaluated by the propagation of realistic model errors to the parameter estimates. Sub-pixel position, sidelobe observations, orbital inaccuracies and ambiguity success rates have been successively addressed. Inaccurate azimuth sub-pixel position can lead to systematic errors in the PS velocities of 0–0.5 mm/year. If acquisitions with high Doppler centroid frequency deviations are included, this error increases to 3 mm/year. Orbital inaccuracies can lead to a systematic errors as well. Deviations in the PS velocities of up to 1 mm/year over 100 kilometers have been shown. Sidelobe observations affect the topographic height estimates and are not independent: they need to be removed if the PS heights are used for PS characterization. Finally, it has been shown that unwrapping success rates of 1 cannot be guaranteed, certainly not in the presence of atmospheric disturbances in areas with a low PS density in the order of 5 PS/km^2.

By means of external validation in the Delft corner reflector experiment, it has been shown that the achievable precision of Envisat double-difference displacements is ~1.6 mm. The precision of leveling double-difference displacements is similar: ~1.5 mm. Moreover, the correlation of Envisat and leveling double-difference displacements is 94%. A slightly lower correlation is obtained between ERS-2 and leveling double-differences (84%), most likely due to the high Doppler centroid frequency deviations in the ERS-2 stack. From this controlled experiment, it can be concluded that the achievable precision of InSAR is in the order of millimeters, which is comparable to the leveling technique.

The estimation of stochastic model parameters is dependent on the redundancy in the system of equations, and can therefore only be performed on the condition that the success rates equal 1, which is not necessarily fulfilled. In the estimation of stochastic model parameters there is a trade-off between the number of parameters that are estimated and their precision. Furthermore, they need to be independent. The Dilution of Precision for PSI has been introduced as a reference independent scalar quality measure for the variance-covariance matrix.

Quality control in PSI does not restrict to a precision and reliability assessment of the technique itself. The concept of idealization precision for deformation modeling has been introduced to describe how well the deformation estimates represent the signal of interest. Different deformation regimes are distinguished: structural instabilities, shallow mass displacements and deep mass displacements. Contrary to traditional geodetic techniques, measurement points in PSI are less well defined. Moreover, the type of reflection (specular, dihedral, trihedral) determines which deformation regimes are present in the measured displacements.

To improve the idealization precision of PSI for deformation monitoring two methodologies have been explained:

- PS characterization by utilizing the topographic height estimates, polarimetric observations and the reflectivity pattern as a function of viewing geometry, and
- statistical methods that either select PS that represent a certain deformation regime (Kriging cross-validation, datasnooping), or decompose PS displacements into components due to different deformation regimes (variance component estimation).

Since it has been shown in Sect. 4.2.4 that ambiguity success rates of 1 cannot be guaranteed, the sensitivity analysis of the PSI parameter estimates for model errors in this chapter represents the most optimistic scenario. It is however possible to create redundancy in PSI deformation monitoring, despite of ambiguity success rates that are not equal to 1. This is achieved by utilizing multiple overlapping independent satellite tracks, which observe the same deformation signal. Hence, redundancy in deformation monitoring is introduced which enables a reliability assessment. The implementation of this multi-track procedure will now be explained in Chap. 5.

Chapter 5
Multi-track PSI

In Chap. 4 it has been demonstrated that the achievable precision of PSI double-difference displacements is the order of several millimeters. Since a precise deformation estimate does not imply a reliable estimate, the influence of model errors has been investigated as well, under the assumption of correct phase unwrapping, see Sect. 4.2. However, ambiguity success rates of 1 cannot be guaranteed, especially not in areas with a low PS density of less than 5 $\mathrm{PS/km^2}$. Hence, an additional assessment of the reliability of the PSI deformation estimates is required.

This chapter introduces multi-track PSI: the same deformation signal is observed from independent overlapping satellite tracks and hence redundancy is introduced in the PSI estimates. The independent overlapping tracks can be formed from different viewing modes (ascending, descending) or from different sensors. Moreover, depending on the latitude of the area of interest, adjacent tracks can be utilized. In the Netherlands, adjacent tracks are overlapping more than 50%. This implies that when combining ascending and descending tracks, the deformation signal of interest is monitored by four independent stacks of observations. Due to the redundancy that is introduced, multiple overlapping tracks can be utilized for a reliability assessment of the PSI deformation estimates.

The PSI estimates of each track have their own coordinate system (datum) that is defined by the acquisition geometry of the master scene. Hence, the multi-track PSI estimates need to be referenced to a common datum, before they can be integrated. In this chapter, a mathematical framework is introduced for the datum connection of multiple tracks, which is simultaneously used to evaluate the reliability of the PSI technique itself. The datum connection procedure consists of two steps:

1. the definition of a unified coordinate system (Ketelaar et al., 2007a),
2. connection of the PSI parameter estimates (Ketelaar et al., 2007b, 2008b).

The datum connection procedure starts with the definition of a unified radar coordinate system that is defined by the viewing geometry of the so-called *master track*. Subsequently, the transformation between the PSI estimates of different tracks (with a different reference PS) is estimated based on the misclosures between the PSI estimates. Due to different viewing geometries, many of the detected PS of different tracks will be physically different. The integration and validation of multi-track PSI estimates is therefore based on the fact that they independently monitor a common deformation signal, although the actual targets may be different.

This chapter starts with the datum connection of PS estimates within a single track in Sect. 5.1. Due to computer memory limitations, the Delft PSI processing is performed using multiple PS networks within a track, and hence requires single-track datum connection. Subsequently, Sect. 5.2 focuses on multi-track datum connection and the reliability assessment of the PSI deformation and height estimates. Finally, Sect. 5.3 shows how the PSI deformation estimates along the satellite line of sight can be spatially decomposed by utilizing the different viewing geometries of the overlapping tracks.

V.B.H. (Gini) Ketelaar, *Satellite Radar Interferometry*,
Remote Sensing and Digital Image Processing 14,

5.1 Single-Track Datum Connection

Depending on the coherence characteristics of the area of interest, the first order PS network can easily contain several tens of thousands of PS candidates. Multiplied by the number of interferometric combinations, this leads to memory problems in solving the system of equations. Therefore, the estimation of the parameters of interest is performed in smaller spatially overlapping PS networks, that are subsequently connected.

In geodesy, datum connection is commonly used for the integration of overlapping point fields, with the purpose to obtain consistent coordinates in the same datum (Teunissen et al., 1987; Baarda, 1981). The transition between two datums can be parameterized as a geometrical transformation, for example as a similarity, orthogonal or affine transformation. To estimate the transformation parameters, identical points are required that are present in both datums.

Datum connection starts with the adjustment of the common points. Subsequently the points outside the overlap (the free points) are corrected based on their correlation with the points in the overlap. As an example, the basic equation for the estimation of a transformation between two point fields reads:

$$E\left\{\begin{bmatrix}\underline{z}\\ \underline{w}\end{bmatrix}\right\}=\begin{bmatrix}I & 0\\ I & A_t\end{bmatrix}\begin{bmatrix}z\\ t\end{bmatrix}, \tag{5.1}$$

where $\underline{z}$ and $\underline{w}$ are the coordinates of the common points in the overlap, and t is the transformation parameter. This model of equations can be reformulated into a system of equations that is based on coordinate differences:

$$E\{\underline{w}-I\underline{z}\}=A_t t. \tag{5.2}$$

If one of the point fields is restricted to remain unchanged, for example in the integration of new observations with benchmarks in an established datum, specific types of connection procedures are applied, such as pseudo least-squares adjustment (ibid.).

A testing procedure can be integrated to indicate the precision and reliability of the point field connection. The procedure can focus on single observations, observations per point, choice and significance of transformation parameters, the variance-covariance matrix and the distribution of the residuals. The free points are observed in only one datum, and can therefore not be tested.

In PSI, the datum connection procedure comprises of both the conversion to a unified coordinate system, and the connection of the PSI parameter estimates. Since the PS in the overlapping networks already refer to the same coordinate system defined by the master scene, the first step in the datum connection procedure (definition of a unified coordinate system) can be omitted. Hence, single-track datum connection is restricted to the connection of PSI estimates of the overlapping networks. These PS networks may all have a different reference PS.

The input in the connection of PSI estimates are the PS displacement (or velocity) and height estimates of the overlapping PS networks, including their variance-covariance matrix. The transformation between the PSI estimates of the different networks is parameterized as a translation. This should be sufficient, since the overlapping area contains the same PSI estimates, that only refer to a different reference PS.

The connection of PS networks is independent of the choice of the reference PS. The relative velocity estimates and their precision remain unchanged. This also holds, when the reference point exhibits non-linear displacements. To demonstrate this, consider overlapping PS networks with displacements d. The reference PS of the displacement networks are a and b, and PS i and j are two common PS in the overlapping part. Reference PS a exhibits non-linear behavior and can be modeled as a second order polynomial, where v is the unknown PS velocity, u the unknown second order component and T^k is the temporal baseline for the kth interferometric combination (the design matrix):

$$E\{\underline{d}_a^k\} = T^k v_a + (T^k)^2 u_a, \tag{5.3}$$

for $k = 1 \dots K$, where K is the number of interferometric combinations. Reference PS b and PS i and j are behaving linear:

$$E\{\underline{d}_b^k\} = T^k v_b; \qquad E\{\underline{d}_i^k\} = T^k v_i; \qquad E\{\underline{d}_j^k\} = T^k v_j. \tag{5.4}$$

The displacements of PS i with reference to a and b can be computed as:

$$E\{\underline{d}_{i(a)}^k\} = T^k v_i - T^k v_a - (T^k)^2 u_a; \qquad E\{\underline{d}_{i(b)}^k\} = T^k v_i - T^k v_b. \tag{5.5}$$

Similarly holds for PS j:

$$E\{\underline{d}_{j(a)}^k\} = T^k v_j - T^k v_a - (T^k)^2 u_a; \qquad E\{\underline{d}_{j(b)}^k\} = T^k v_j - T^k v_b. \tag{5.6}$$

The relative displacements between PS i and PS j are the same in both networks, regardless of the reference PS:

$$\begin{aligned} E\{\underline{d}_{ij(a)}^k\} &= E\{\underline{d}_{j(a)}^k\} - E\{\underline{d}_{i(a)}^k\} = T^k v_j - T^k v_i, \\ E\{\underline{d}_{ij(b)}^k\} &= E\{\underline{d}_{j(b)}^k\} - E\{\underline{d}_{i(b)}^k\} = T^k v_j - T^k v_i. \end{aligned} \tag{5.7}$$

If the displacement series are correctly unwrapped, the nonlinear part of the reference point will cancel in the relative velocities between the common points in the overlap. This means that the relative velocity differences in the overlap are the same irrespective of the displacement behavior of the reference points.

For the overlapping PS networks within a single track, PSI parameter estimation including the estimation of atmospheric phase screens and phase unwrapping are performed independently. After application of the datum connection, the PSI estimates should agree within the limits that are defined by the precision of the PSI parameter estimates and the potential model errors, see Chap. 4. The application of single-track datum connection for the tracks that cover the Groningen area is addressed in Sect. 6.3. There, the misclosures are evaluated in the overlap of four PS networks that cover the entire scene.

5.2 Multi-track Datum Connection

Systematic effects in the PSI parameter estimates due to residual errors that propagate over a large spatial extent are difficult to identify within a single stack, certainly when the success rates of phase unwrapping and the APS accuracy decrease in rural areas with a low PS density. The combination of multiple independent tracks introduces redundancy in the estimation of the deformation signal and hence enables a reliability assessment.

The integration of PSI estimates from multiple tracks is considered as a datum connection problem. The PSI estimates of independent tracks are located in their own local radar coordinate datum and refer to their own reference PS. As a consequence, a datum connection between all tracks needs to be established. This datum connection consists of two steps:

1. the definition of a unified coordinate system,
2. connection of the PSI parameter estimates (displacements, heights).

In the datum connection procedure, the deformation estimates from multiple independent tracks are integrated. Therefore, the likelihood that PS from different tracks refer to the same physical target or deformation regime should be optimized. In other words, the multi-track PS coordinates should be unambiguous. The location of a PS can either be expressed in the radar coordinate system or in a geographic coordinate system. Georeferencing each track independently does not resolve for the range and azimuth timing error per track and the uncertainty in the reference PS height. Hence, it is proposed to define a common radar coordinate datum: the acquisition geometry of one of the tracks that is appointed as *master track*. All other tracks are referred to as slave tracks.

The conversion of multiple overlapping tracks to a common radar coordinate datum leads to a consistent PS localization. Furthermore, it reduces the degrees of freedom in the georeferencing to one range and azimuth timing error. Moreover, the uncertainty in reference PS heights decreases in the comparison of multiple tracks, prior to georeferencing.

5.2.1 *Unified Radar Datum*

This section describes the transformation of the local radar coordinates per track to a unified radar datum. It starts with the estimation of transformation parameters based on precise orbits, that are subsequently refined by utilizing the PS point fields or the multi-image reflectivity maps.

5.2.1.1 Orbit Based Radar Coordinate Transformation

Based on the precise orbits of the master scenes for each track, an approximate radar coordinate transformation between the tracks can be estimated. For T tracks,

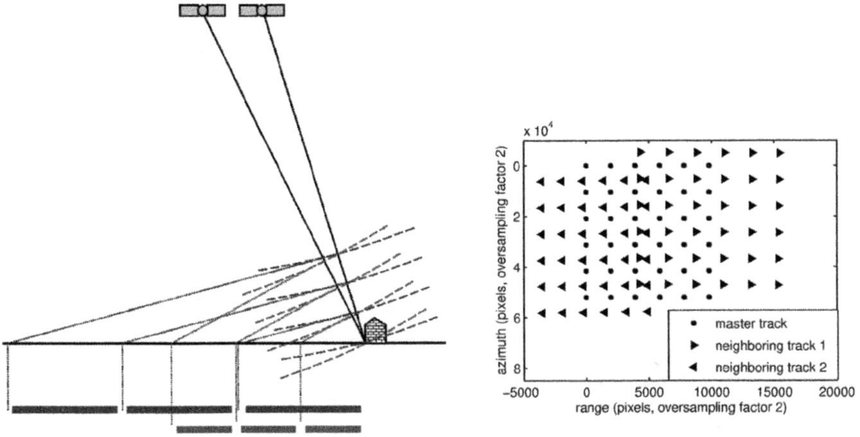

Fig. 5.1 A different incidence angle of adjacent tracks results in a different ground resolution (left). The effect of this difference in resolution is visible in the projection of a set of equally spaced radar coordinates in the slave track to the radar datum of the master track (right)

$T-1$ independent transformations can be defined. If the ground level height differences are limited (<50 meters), relative pixel location errors are in the order of 0.1 pixel for a baseline of 50 kilometers (between adjacent tracks). This falls within the coregistration precision. Hence, uniformly spaced subsets of radar coordinates in the master track reference datum are transformed to geographical WGS84 coordinates on the ellipsoid (see Fig. 5.1), that are subsequently projected to the radar datum of a neighboring (overlapping) track. This results in two sets of radar coordinates that refer to the same geographical locations.

If the ground level height differences in the area of interest exceed ~50 meters, the same procedure can be followed, but the ellipsoidal heights have to be utilized in the transformation to and from the geographic locations. One option is to obtain the heights from an external DEM. The other option is to utilize the estimated PS heights. Since PS height estimates are relative to a reference PS, the ellipsoidal height of this reference PS needs to be obtained.

The offsets between the radar coordinates in the master and slave track of the points on the ellipsoid are regarded as the observations in the system of equations for the estimation of the radar coordinate transformation between the tracks. The radar coordinate transformation is parameterized as a pth degree polynomial:

$$\begin{aligned}\Delta\xi(\xi,\eta) &= \sum_{i=0}^{p}\sum_{j=0}^{i}\alpha_{i-j,j}\xi^{i-j}\eta^{j},\\ \Delta\eta(\xi,\eta) &= \sum_{i=0}^{p}\sum_{j=0}^{i}\beta_{i-j,j}\xi^{i-j}\eta^{j},\end{aligned} \tag{5.8}$$

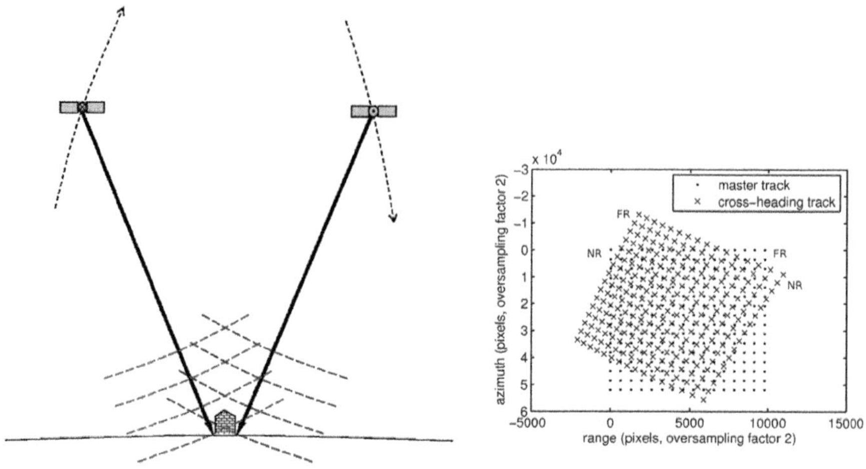

Fig. 5.2 The input for the radar coordinate transformation are two sets of radar coordinates in the datum of respectively the master and the slave track (left). The satellite direction is normal to the plane of the paper. The different viewing geometry of overlapping tracks can be visualized in the radar datum of the master track (right). Here, the relative distortion of cross-heading tracks is depicted

Table 5.1 Precision of the transformation from slave track radar datum to master track radar datum for different polynomial degrees. The maximum range coordinate residuals (pixels, oversampling factor 2) are listed for the points that have been used in the estimation of the transformation parameters

polynomial degree	adjacent	cross-heading
2	3	40
3	0.3	6
4	0.15	1.5
5	0.1	0.25

where $\Delta\xi$ and $\Delta\eta$ are the offsets in respectively azimuth and range direction. The parameters α and β describe the transformation from the radar datum of the slave track to the radar datum of the master track. This transformation can be estimated for any pair of overlapping tracks, both adjacent and cross-heading tracks, see Fig. 5.2.

The estimation procedure of the radar coordinate transformation parameters is followed by a testing procedure that evaluates the residuals over the full overlap between two tracks. Depending on the precise orbits and the relative distortion of the radar coordinate system, the degree of the polynomial has to be increased to ensure that the transformation is at sub-pixel level. Table 5.1 shows the results of a case study for the Groningen area (see Sect. 6.4) that indicates that a polynomial degree of 5 was required to ensure a radar coordinate transformation at sub-pixel level both for adjacent and cross-heading tracks.

5.2.1.2 Matching Using PS Point Fields

The orbit based multi-track radar coordinate transformations do not account for timing errors in range and azimuth direction. These have to be estimated based on the image contents. Under the null hypothesis, timing errors are parameterized as an offset in range and azimuth direction.

PS targets are mainly located at man-made features and structures in the terrain that do not suffer from temporal decorrelation. Hence, PS in overlapping tracks are located in the same urban(ized) areas, although the reflection origin and type may not be the same. Since similar PS point fields are available for each track, the optimal shift between their location fields can be estimated, even for cross-heading tracks, see Fig. 5.3. Figure 5.4 shows the search space of range and azimuth shifts for the optimal match between PS point fields of different tracks. Compared to the situation for cross-heading tracks, the maximum of corresponding PS that indicates the optimal range and azimuth shift is more evident for adjacent tracks. Still, the estimation of the range and azimuth shift is sub-optimal, due to the presence of local maxima. Hence, additional information will be utilized for the estimation of the range and azimuth shift: the multi-reflectivity maps.

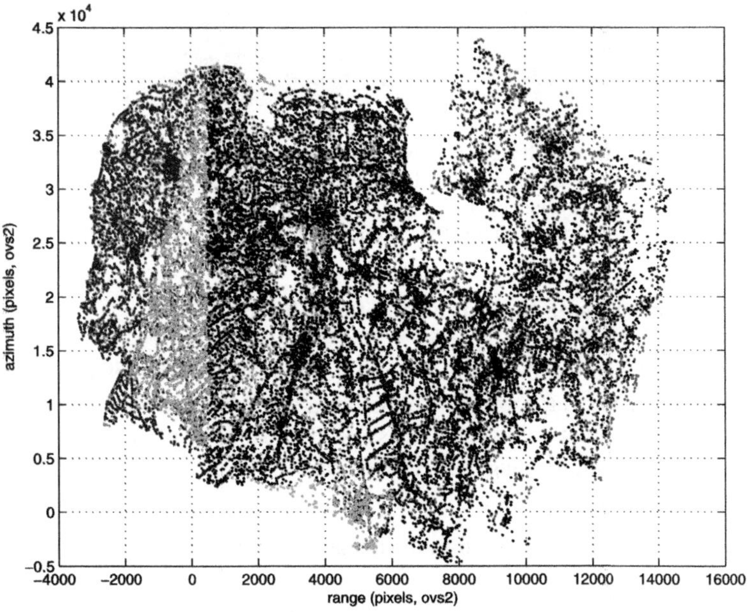

Fig. 5.3 PS point fields of six tracks in a unified radar datum: the PS locations correspond with the buildings and structures in the terrain

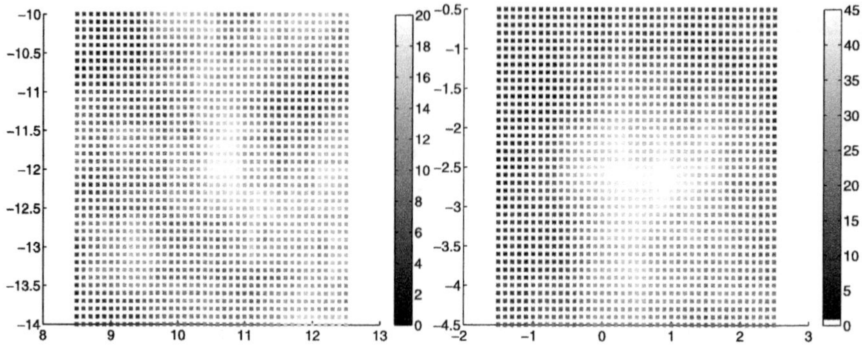

Fig. 5.4 Datum connection based on PS point fields: search space for the optimal range and azimuth shift for (a) two cross-heading tracks and (b) two adjacent tracks. The optimal range and azimuth shift correspond with the shifts that maximize the number of corresponding PS locations

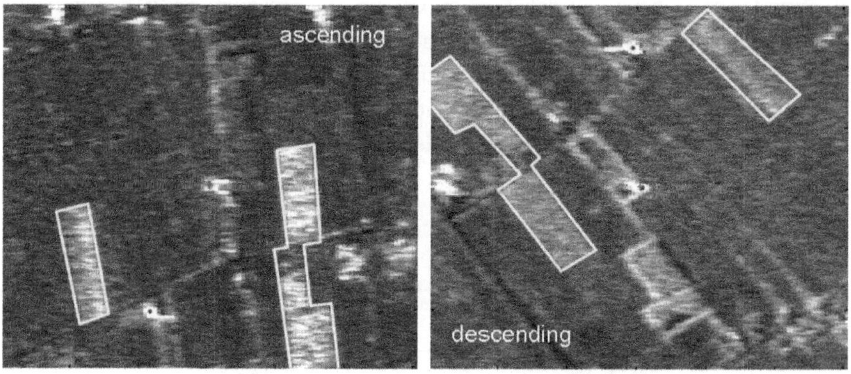

Fig. 5.5 Cross-heading multi-image reflectivity windows. Due to the different viewing geometry, the windows are rotated with respect to each other and the ground resolution is slightly different. Despite of the different viewing geometry, reflections that refer to the same structure/building can be observed (dots). Furthermore, the spatial pattern of fields in this rural area can be recognized

5.2.1.3 Matching Using Multi-image Reflectivity Maps

Besides the PS locations, the multi-image reflectivity maps per track can be utilized to estimate a refined radar coordinate transformation between tracks. As the viewing geometry from different tracks is not the same, the ground resolution and orientation of the overlapping multi-image reflectivity patches vary, see Fig. 5.5. Hence, the following procedure is applied:

1. selection of evenly distributed multi-image reflectivity windows at locations of high amplitude,
2. resampling of the selected windows to the master track radar coordinate system using the initial orbit based transformation,
3. estimation of the range and azimuth shift using correlation optimization.

Matching using multi-image reflectivity maps is incoherent matching, contrary to the coregistration of SAR scenes, which is based on the complex coherence of the coregistration windows.

In stead of using a search space for the determination of the range and azimuth shift, the application of an image matching technique that is used in digital photogrammetry has been investigated (Gruen and Baltsavias, 1985). The multi-image reflectivity windows of different tracks can be seen as geometrically transformed patches of amplitude observations, see Fig 5.5. If the height differences in multi-track multi-image reflectivity windows are less than ~50 meters, the transformation between the patches can be locally approximated by a similarity or affine transformation, see Fig. 5.2.

By means of the initial orbit based transformation, the multi-image reflectivity windows of the slave track can already be approximately resampled to the radar datum of the master track. The range and azimuth shift are the only parameters that are additionally required to identify common targets in the master track and the slave track. Hence, the transformation between the multi-image reflectivity window in the master track and its corresponding resampled window in the slave track is parameterized as a translation in range and azimuth direction. The equivalent of the matching equations from photogrammetry (ibid.) would then read:

$$E\{\underline{f}(\xi,\eta) - \underline{g}^0(\xi,\eta)\} = \frac{\delta g^0(\xi,\eta)}{\delta\xi}d\xi + \frac{\delta g^0(\xi,\eta)}{\delta\eta}d\eta, \tag{5.9}$$

where:

ξ	azimuth coordinate,
η	range coordinate,
$\underline{f}(\xi,\eta)$	amplitude observations in the multi-image reflectivity window of the master track,
$\underline{g}^0(\xi,\eta)$	amplitude observations in the multi-image reflectivity window of the slave track,
$\frac{\delta g^0(\xi,\eta)}{\delta\xi}, \frac{\delta g^0(\xi,\eta)}{\delta\eta}$	amplitude gradients in azimuth and range direction,
$d\xi, d\eta$	differential azimuth and range coordinates.

$d\xi$ and $d\eta$ can be optionally parameterized as a higher degree polynomial. The system of equations is linearized and solutions are obtained in an iterative way. The advantage of this method is that it guides the estimation of the transformation parameters; it is not necessary to explore the entire search space of potential range and azimuth shifts. The disadvantage of the application of this method to radar data is that the sensitivity of the amplitude observations for the radar viewing geometry decreases the robustness of the image matching technique.

Finally, it has been chosen to estimate the range and azimuth shifts by means of correlation optimization. Optionally, the image based matching technique has been additionally applied to further increase the precision of the range and azimuth shifts. The obtained precision is 0.25 pixel for adjacent tracks, and 0.5 pixel for cross-heading tracks. This is sufficient to identify multi-track PS within resolution cell distance.

5.2.2 Connection of PSI Estimates

Now the locations of all PS have been defined in the radar datum of the master track, we can proceed to the second step of the datum connection procedure: the connection of the multi-track PSI estimates.

5.2.2.1 Clusters of Multi-track Identical and Neighboring PS

The transformation of PS coordinates to the radar datum of the master track enables the identification of physically identical scatterers and clusters of neighboring scatterers in adjacent and even cross-heading tracks, see Fig. 5.6. The incidence angle of adjacent tracks differs only a few degrees, implying that dihedral and trihedral targets that point in the satellite look direction are likely to be observed from multiple tracks. In contrast, cross-heading tracks have an nearly opposite viewing direction. Cylindrical poles for example are observed in ascending and descending mode (Perissin, 2006), but such scatterers do not occur frequently in a rural area.

We assume that physically identical targets represent the same deformation regime, provided that the reflection type (specular, dihedral) is the same. This does not imply that only identical scatterers can be used for the datum connection. The detected PS in overlapping tracks clearly follow man-made structures in the terrain, as shown in Fig. 5.7. This allows grouping of nearby PS, which displacements are potentially caused by the same deformation regime(s). The shorter the PS distance, the higher the likelihood that they represent the same deformation regime(s).

5.2.2.2 Multi-track System of Equations

The multi-track PS clusters are utilized to compute the transformation parameters that perform the integration of the multi-track PSI estimates (displacements, heights). Since the PS within a cluster refer to identical targets or are at a close distance, their displacement estimates are assumed to represent the same deformation regime, apart from a translation due to a different reference PS. One option for the integration of PSI displacement estimates is a joint estimation of the deformation signal. However, this requires a-priori knowledge on the functional modeling of (unknown) deformation regimes. Another option is to utilize the *misclosures* within a PS cluster as observations in the connection of PSI estimates. Here, it is not necessary to attribute the PS clusters to a certain deformation regime: the displacements in different tracks only differ due to a different reference PS, irrespective of the deformation regime(s).

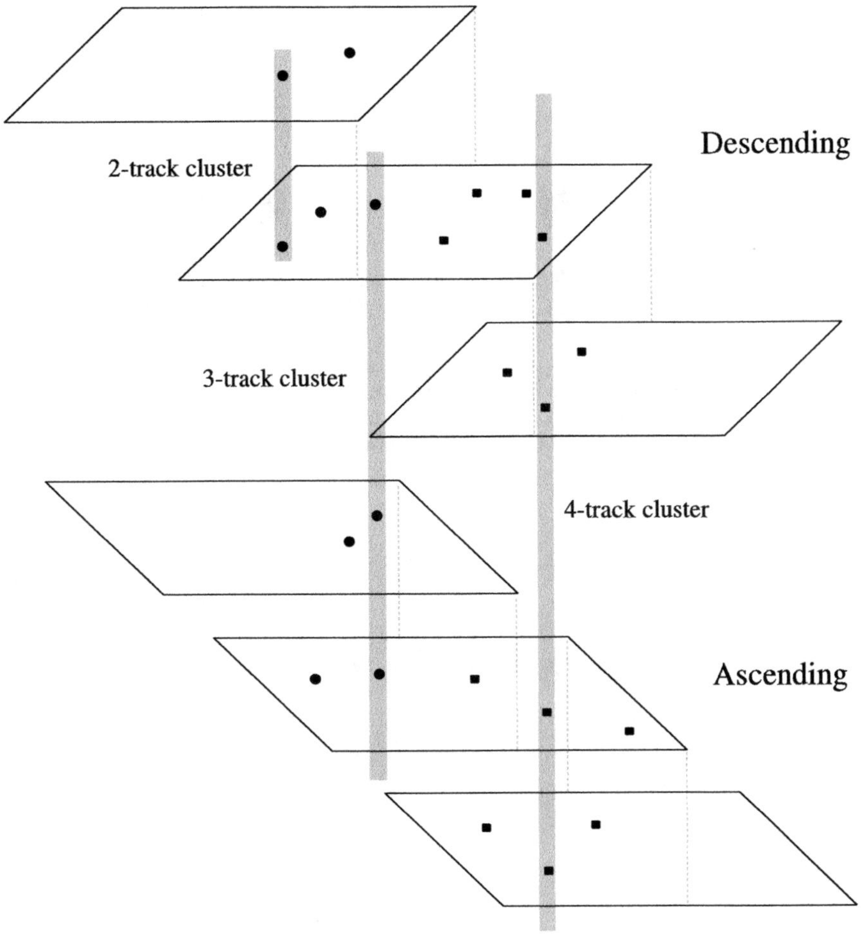

Fig. 5.6 Schematic overview of PS clusters in overlapping adjacent and cross-heading tracks. The vertical gray lines represent the PS clusters: groups of PS from different tracks that have (approximately) the same coordinates in the radar datum of the master track

If a PS cluster consists of PS from T tracks, only $T-1$ independent observations can be formed for this cluster. As an example, suppose the area of interest has been observed by four tracks $T_1 \ldots T_4$. Furthermore, suppose that M clusters of neighboring PS can be formed that are observed by all four tracks. The multi-track PS estimates describe the same deformation regime but refer to a different reference PS. The PSI parameter estimates (displacements, heights) should theoretically only be subject to an unknown translation (bias) because of a different reference PS. Hence, under the null hypothesis of a translation between the estimated PS velocities of the tracks, the functional model

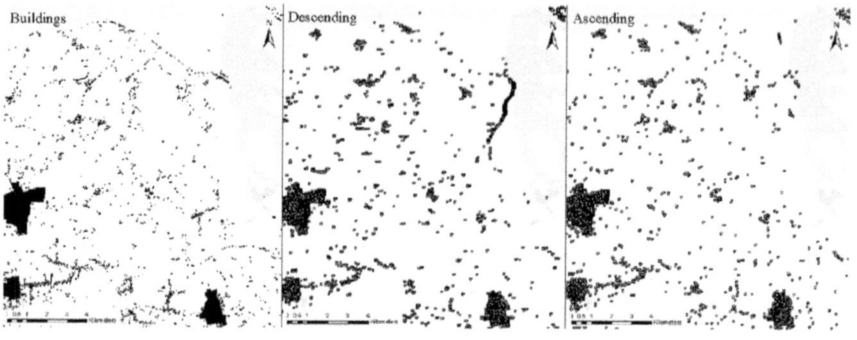

Fig. 5.7 Distribution of buildings and urban(ized) areas in the northern part of the Netherlands (left, source: © Kadaster en de Openbare Registers, Apeldoorn) and geocoded PS from overlapping tracks (middle, right). Middle: PS from two descending tracks. Right: PS from two ascending tracks. It can be seen that the PS coincide with man-made structures in the terrain

reads:

$$E\{\underline{t}\} = Ax, \tag{5.10}$$

$$\begin{bmatrix} I_M & -I_M & 0 & 0 \\ I_M & 0 & -I_M & 0 \\ I_M & 0 & 0 & -I_M \end{bmatrix} \begin{bmatrix} \underline{v}^{T_1}_{1\ldots M} \\ \underline{v}^{T_2}_{1\ldots M} \\ \underline{v}^{T_3}_{1\ldots M} \\ \underline{v}^{T_4}_{1\ldots M} \end{bmatrix} = I_3 \otimes e_M \begin{bmatrix} t_0^{T_{12}} \\ t_0^{T_{13}} \\ t_0^{T_{14}} \end{bmatrix}, \tag{5.11}$$

where:

$\underline{t}$	vector of misclosures,
$\underline{v}$	estimated PS velocity,
t_0	translation parameter,
$\otimes$	Kronecker product,
I	identity matrix,
e_M	$M \times 1$ vector with all ones.

The corresponding variance-covariance matrix of the input observations (the PSI velocity estimates) reads:

$$Q_t = \begin{bmatrix} Q_v^{T_1} & 0 & 0 & 0 \\ 0 & Q_v^{T_2} & 0 & 0 \\ 0 & 0 & Q_v^{T_3} & 0 \\ 0 & 0 & 0 & Q_v^{T_4} \end{bmatrix}, \tag{5.12}$$

where $Q_v^{T_i}$ is variance-covariance matrix of the velocity estimates for track T_i. Due to the double-difference combinations, $Q_v^{T_i}$ is a full matrix. Due to the linear combinations in (5.11), the variance-covariance matrix of the velocity estimate misclosures is a full matrix as well, although the PSI estimates from different tracks are uncorrelated.

If the PSI parameter estimations for a single track are performed per arc, the stochastic model is restricted to the variance-covariance matrix of the parameter estimates (displacements, heights) per arc. The interpretation of either the coherence or the displacement variance as an absolute precision measure leads to the misleading conclusion that PS further away from the reference PS have a lower precision. A solution in the situation where covariances between the PSI parameter estimates are not available is to reconstruct a substitute variance-covariance matrix. This is performed based on the variance-covariance matrix of double-differences observations Q_y, see e.g., (4.19), and the design matrix A, see (3.12):

$$Q_{\hat{x}} = (A^T Q_y^{-1} A)^{-1}. \tag{5.13}$$

The system of observation equations from (5.11) can be formulated as a system of condition equation as well (Teunissen, 2000a). The system of condition equations has the advantage that the variance-covariance matrix of the velocity estimate misclosures does not need to be inverted to obtain the translation parameter estimates. The design matrix of the model of observation equations and the model of condition equations are denoted as A and B respectively. The multiplication of both design matrices equals the zero matrix:

$$B^T A = 0. \tag{5.14}$$

For (5.11) the B matrix would read:

$$B^T A = \begin{bmatrix} e_{M-1} & -I_{M-1} & 0 & 0 & 0 & 0 \\ 0 & 0 & e_{M-1} & -I_{M-1} & 0 & 0 \\ 0 & 0 & 0 & 0 & e_{M-1} & -I_{M-1} \end{bmatrix} \begin{bmatrix} e_M & 0 & 0 \\ 0 & e_M & 0 \\ 0 & 0 & e_M \end{bmatrix} = 0, \tag{5.15}$$

where B has the dimension $(3 \cdot M) \times (3 \cdot (M-1))$ and A has the dimension $(3 \cdot M) \times 3$, for four tracks.

5.2.2.3 Step-wise Estimation Based on PS Distance

When PS clusters with increasing mutual PS distance (see Fig. 5.8) are included in the connection of deformation estimates, the dimension of the system of equations of (5.10) grows rapidly. Hence, this section investigates if step-wise (static recursive) estimation of the transformation parameters in the connection procedure can be applied. Step-wise estimation of the transformation parameters would reduce the dimensions of the system of equations, and hence the transformation parameters could be obtained relatively quick. If both the estimates and their variance-covariance matrix are taken into account in each recursive step, the step-wise solution equals the solution obtained in batch mode (Teunissen, 2001a).

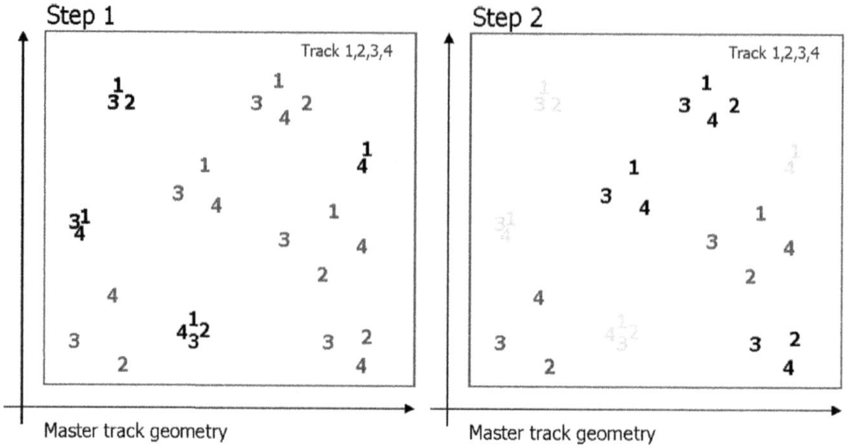

Fig. 5.8 Step-wise estimation of transformation parameters using clusters of identical or neighboring PS from multiple tracks. In each recursive step, clusters with a larger mutual distance are included

In each step k the following model is solved:

$$E\left\{\begin{bmatrix}\hat{\underline{x}}_{(k-1)} \\ \underline{y}_k\end{bmatrix}\right\} = \begin{bmatrix} I \\ A_k \end{bmatrix} x; \quad D\left\{\begin{bmatrix}\hat{\underline{x}}_{(k-1)} \\ \underline{y}_k\end{bmatrix}\right\} = \begin{bmatrix} Q_{\hat{x}(k-1)} & 0 \\ 0 & Q_k \end{bmatrix}, \tag{5.16}$$

where $\hat{\underline{x}}_{(k-1)}$ are the transformation parameters obtained in the previous step with variance-covariance matrix $Q_{\hat{x}_{(k-1)}}$. The vector $\underline{y}_k$ contains the PS cluster misclosures with variance-covariance matrix Q_k. The step-wise connection can be performed in two ways:

1. consecutive connecting of (parts of) the tracks, or
2. connection of all tracks simultaneously using PS clusters of increasing mutual distance.

Since the connection procedure is based on observations from different tracks that are assumed to represent the same deformation signal, the second option is chosen. The step-wise estimation is initialized by the PS clusters within resolution cell distance. In each subsequent step, the observations of PS clusters with a larger mutual distance are added. All PS are uniquely addressed to only one cluster. Datasnooping is applied to remove PS cluster observations that do not refer to the same deformation regime.

A condition on the application of the step-wise procedure is that the partitions of PS observations added in each step should be uncorrelated. Only for the connection of entire tracks this condition is valid. Hence, strictly spoken, step-wise connection cannot be applied. Further research will have to address the practical implications of considering the PS clusters in subsequent steps as uncorrelated.

5.2.3 Spatial Trends

If the PS parameter estimates in the individual tracks would be free of model inaccuracies, it is sufficient to model PSI parameter estimate misclosures between tracks by a translation (bias). However, due to possible residual orbital, atmospheric and unwrapping errors, an alternative hypothesis is evaluated that includes a spatial trend in range and azimuth direction:

$$H_A : \begin{bmatrix} I & -I \end{bmatrix} \begin{bmatrix} \underline{v}^{T_1} \\ \underline{v}^{T_2} \end{bmatrix} = \begin{bmatrix} \xi & \eta & 1 \end{bmatrix} \begin{bmatrix} t_\xi^{T_{12}} \\ t_\eta^{T_{12}} \\ t_0^{T_{12}} \end{bmatrix}, \tag{5.17}$$

where t_ξ and t_η are the azimuth and range dependent transformation parameters, see Fig. 5.9.

The hypotheses are evaluated by means of the overall model test and the individual $\underline{w}$-teststatistics (Teunissen, 2000b). $\underline{w}$-teststatistics are usually applied to trace outliers in individual observations, but also their distribution can also be used to trace systematic effects. If the functional and stochastic model are correct, the $\underline{w}$-teststatistics have a standard normal distribution. Consequently, deviations from this standard normal distribution indicate possible unmodeled effects. Furthermore, the misclosures after datum connection are evaluated, and should theoretically be within the range of their precision from the single track estimation.

After datum connection, the PSI results are mutually consistent in the reference system of the master track. The precision of the estimated transformation parameters follows from the least-squares adjustment based on (5.11). However, although the transformation parameters may have a high precision, the reference system of the master track itself can still contain small systematic components due to possible residual orbital, atmospheric and unwrapping errors. These cannot be unambiguously addressed to either real deformation signal or unmodeled residual components in the PSI estimation. By utilizing the large spatial extent that is covered by a track and a-priori assumptions on non-deforming areas, the spatial trend in the reference track can be estimated and removed. This is justified, as long as the corrections and the error bounds are clearly defined.

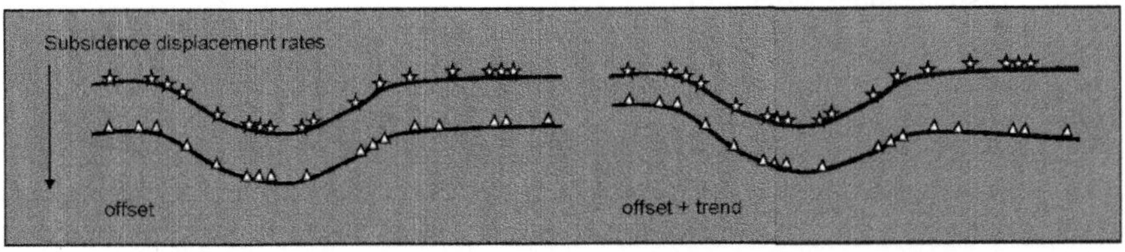

Fig. 5.9 Multi-track offset (null hypothesis) or offset plus trend (alternative hypothesis)

5.3 Decomposition of Line of Sight Deformation

Because the deformation estimates of independent tracks observe the same deformation signal from different viewing geometries, the line of sight (LOS) displacements can be further decomposed into horizontal and vertical displacements. If the deformation signal of interest includes a horizontal component and is observed from only one track, a-priori knowledge on the deformation mechanism is required. It two tracks are available, preferably an ascending and a descending track, the vertical component and one horizontal component can be directly computed from the PSI estimates. This would also aid the discrimination between different deformation regimes: shallow compaction may be dominated by vertical displacements, whereas subsidence due to gas and oil extraction exhibits horizontal displacements as well.

5.3.1 System of Equations

The multi-track PS clusters from the datum connection are utilized for a further decomposition into horizontal and vertical displacements (Hanssen, 2001). Each PS cluster should contain at least two and maximum four PS originating from different viewing geometries (ascending, descending, and adjacent tracks). For a cluster of two multi-track PS, a decomposition into the vertical component and one horizontal component along a certain viewing direction is possible. For a cluster of three multi-track PS theoretically a decomposition into vertical, east and north components is possible. Finally, if a cluster consists of four multi-track PS, this decomposition can even be performed redundantly. However, due to the acquisition geometry, the north component can only be obtained with a low precision for a single PS cluster (ibid.).

The system of equations for the decomposition of PS deformation estimates into the vertical component and a horizontal component in the ascending look direction reads:

$$E\left\{\begin{bmatrix} \underline{d}_r^{\text{asc}} \\ \underline{d}_r^{\text{desc}} \\ \underline{d}_r^{\text{asc(a)}} \\ \underline{d}_r^{\text{desc(a)}} \end{bmatrix}\right\} = \begin{bmatrix} \cos\theta_i^{\text{asc}} & -\sin\theta_i^{\text{asc}} \\ \cos\theta_i^{\text{desc}} & -\sin\theta_i^{\text{desc}}\cdot\cos\Delta\alpha \\ \cos\theta_i^{\text{asc(a)}} & -\sin\theta_i^{\text{asc(a)}} \\ \cos\theta_i^{\text{desc(a)}} & -\sin\theta_i^{\text{desc(a)}}\cdot\cos\Delta\alpha \end{bmatrix} \begin{bmatrix} d_u \\ d_{h_{\text{ald}}} \end{bmatrix}, \tag{5.18}$$

with:

d_r displacement along LOS,
d_u vertical displacement,
$d_{h_{\text{ald}}}$ projection of horizontal displacement in azimuth look direction,
(a) adjacent track.

The incidence angle θ_i is determined from near range to far range with reference to the WGS84 ellipsoid. The heading of the satellite is computed from the orbits of the master acquisitions for each track. The satellite heading difference between ascending and descending mode is denoted by $\Delta\alpha$.

5.3.2 *Quadtree Decomposition*

If the deformation signal of interest covers a large spatial extent and is spatially correlated (smooth), the redundancy in the displacement decomposition can be increased by merging neighboring multi-track PS clusters. This implies that for a certain spatial area all multi-track PSI estimates are utilized to estimate one vertical component and one horizontal component. It is based on the assumption that the vertical and horizontal component can be considered approximately constant over a certain area within the accuracy range of the multi-track PSI estimates. An increased redundancy results in a higher precision of the spatial decomposition estimates.

The grouping of multi-track PS clusters for spatial decomposition can be based on a quadtree decomposition of the deformation signal. The quadtree decomposition starts with the subdivision of the deformation signal into square blocks of equal size. Each block can be further subdivided, depending on a teststatistic that represents the homogeneity of the deformation signal within that block. For example: a block is further subdivided is the difference between the minimum and maximum PS velocity exceeds 1 mm/year. Figure 4.29 shows an example of a quadtree decomposition of a deformation signal. On the slopes of the subsidence bowls, where the displacement rates change more rapidly, the quadtree decomposition is more detailed. On the contrary, in stable areas, blocks have not been subdivided.

5.4 Conclusions

A mathematical framework for the datum connection of multiple independent overlapping tracks has been newly developed. Based on orbits, PS point fields, and the multi-image reflectivity maps per track, it has been demonstrated that multi-track PS locations can be converted into a common radar datum defined by the *master track*. Subsequently, clusters of identical or neighboring PS from different tracks are detected. Based on the multi-track misclosures, transformation parameters are estimated for the datum connection of PSI estimates. After the datum connection of multiple independent overlapping tracks, misclosures between the tracks can be analyzed and hence a statement on the reliability of the results can be given. Furthermore, the deformation estimates can be further decomposed into vertical and horizontal components based on the different viewing geometry of the tracks. A quadtree decomposition of the signal of interest can be applied to increase redundancy in the estimation of the spatial components.

By including the multi-track connection procedure in PSI quality control, both the precision and the reliability of the PSI technique can be assessed. Chapter 6 will now demonstrate the applicability of PSI for monitoring subsidence due to hydrocarbon production in the Groningen area.

Chapter 6
PSI Subsidence Monitoring in Groningen

In this chapter, the concepts of Chaps. 3, 4 and 5 are applied for monitoring subsidence due to hydrocarbon production. The focus lies on the gas fields in the north-eastern part of the Netherlands, in particular on the Groningen gas field. It has a diameter of ~30 kilometers and is covered by six ERS and Envisat tracks. These tracks—each with a footprint of ~100 × 100 km—cover together almost the entire northern part of the Netherlands and a part of Germany, see Fig. 6.1. Due to the latitude of the Groningen gas field (53°), the adjacent tracks overlap approximately 50%. The collected ERS SAR images have been acquired in image mode with VV polarization since 1992. Envisat has started data acquisition in this mode in 2003.

This chapter starts with the application of InSAR processing and the Delft PSI parameter estimation (DePSI) for this specific study in Sect. 6.1. Subsequently, the ERS and Envisat PSI deformation and height estimates will be presented in Sect. 6.2. The precision of the PSI estimates is investigated in Sect. 6.3. The overlap between PS networks in single tracks is utilized to cross-check the precision measures. The effect of unmodeled residual components is addressed as well: a large scale spatial trend is clearly visible in the Envisat PS velocity estimates. The potential causes of this trend are investigated, and the strategy to remove the trend. After addressing the precision of the PSI estimates and potential model errors, the multi-track procedure is applied for a reliability assessment in Sect. 6.4. When the precision and reliability of PSI as a measurement technique have been quantified, Sect. 6.5 focuses on increasing the idealization precision of PSI for monitoring subsidence due to hydrocarbon production. A-priori knowledge on the spatio-temporal behavior of the signal of interest will be utilized. Furthermore, the available techniques for PS characterization are investigated. The purpose of this chapter is to present PSI deformation estimates for monitoring subsidence due to hydrocarbon production, including a description of their (idealization) precision and reliability.

6.1 InSAR Processing Strategy

The PSI analyses over the Groningen area have been performed using the Delft Object-oriented Radar Interferometric software (Doris (Kampes and Usai, 1999)) and the Delft implementation of PSI (DePSI). This section explains how these concepts (see Sects. 3.1 and 3.4) have been applied in practice.

6.1.1 Data Coverage and Master Selection

The ERS and Envisat tracks that cover the Groningen area are listed in Table 6.1. Each track has been processed using interferometric combinations that refer to

V.B.H. (Gini) Ketelaar, *Satellite Radar Interferometry*,
Remote Sensing and Digital Image Processing 14,

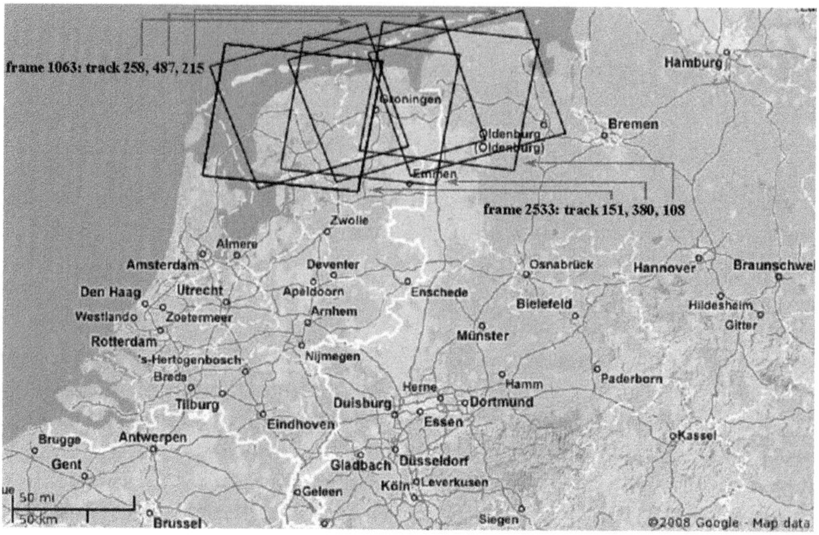

Fig. 6.1 Spatial coverage of the six ERS tracks that cover the Groningen subsidence area (map: © 2008 Google – Map data © 2008 Tele Atlas)

Table 6.1 The six ERS tracks and one Envisat track that cover the Groningen subsidence area. In total, these tracks cover the entire northeastern part of the Netherlands and a part of Germany. The number of scenes refers to the selection in which acquisitions with a Doppler deviation >500 Hz from the master's Doppler centroid frequency have been excluded. The master has been selected based on the stack coherence and the geographic ground coverage

sensor	track	frame	mode	location		#scenes	master
ERS	151	2533	desc	Friesland	(west)	75	21-03-1997
ERS	380	2533	desc	Groningen	(main)	73	20-07-1997
ERS	108	2533	desc	Germany	(east)	63	05-08-1997
ERS	258	1063	asc	Friesland	(west	32	06-06-1997
ERS	487	1063	asc	Groningen	(main)	33	27-06-1999
ERS	215	1063	asc	Germany	(east)	25	14-01-1997
Envisat	380	2533	desc	Groningen	(main)	41	29-05-2005

a common master. The master acquisitions have been selected based on the stack coherence, see (3.8). Additionally, since entire scenes have been processed, the common ground coverage of the acquisitions within a stack has been considered as well, based on the geographic location of the acquisitions. The location of the majority of the acquisitions in a stack varies within ~0.02 degrees. Only the scenes with a deviating Doppler centroid frequency have a significant shift in ground coverage, see Fig. 6.2.

Before proceeding with the master images that were selected on stack coherence and ground coverage, it is important to verify if the reflectivity of the earth surface at the acquisition time was not affected by weather conditions. Hence, the daily

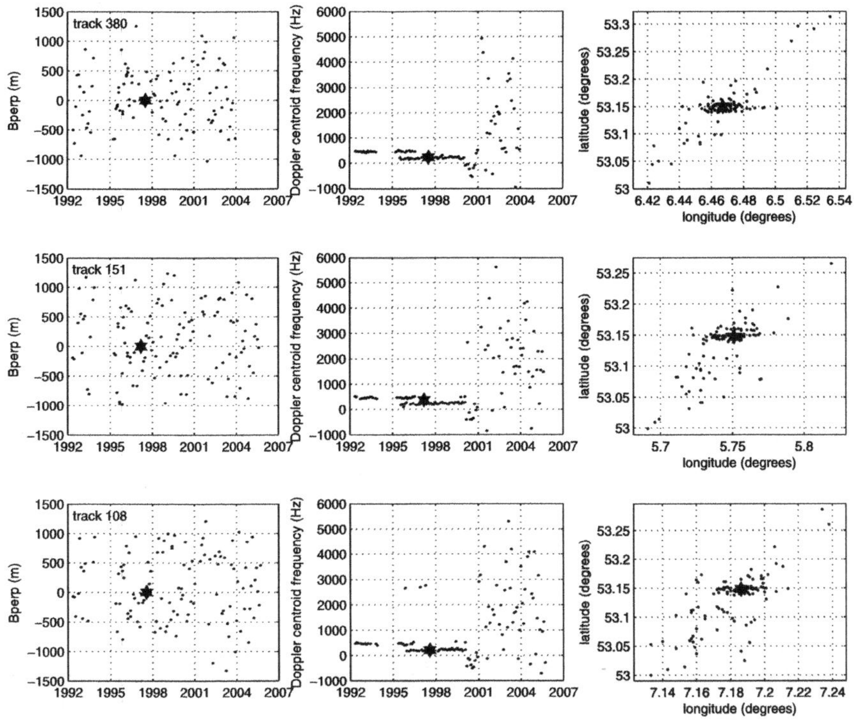

Fig. 6.2 Acquisition geometry of the descending ERS tracks: temporal baseline versus perpendicular baseline (left), temporal baseline versus Doppler centroid frequency (middle) and geographic location of the center of the scenes (right)

weather conditions of the master acquisitions have been checked on the website of KODAC (2008). Especially for track 215, which master acquisition took place in January, it is important to know that the earth surface was not covered with snow or ice.

The acquisition geometries per track (perpendicular baseline and Doppler centroid frequency as a function of temporal baseline) are shown in Figs. 6.2, 6.3, and 6.4. Not all acquisitions have been included in the PSI processing. The observations of ERS images with deviating Doppler centroid frequencies need to be corrected for their sub-pixel position (Marinkovic et al., 2006), which imposes an additional uncertainty on the investigation of the PSI applicability in a rural area with small subsidence rates. Moreover, Cassee (2004) has noticed that the amplitude observations of acquisitions with a deviating Doppler centroid frequency deteriorate the PS selection based on normalized amplitude dispersion. Since the goal of this research is to prove the applicability of PSI for subsidence monitoring in the Groningen area, it has been chosen to exclude the acquisitions with a Doppler frequency difference larger than 500 Hz from the master's Doppler centroid frequency. The maximum effect on the estimated displacement of a Doppler difference of 500 Hz

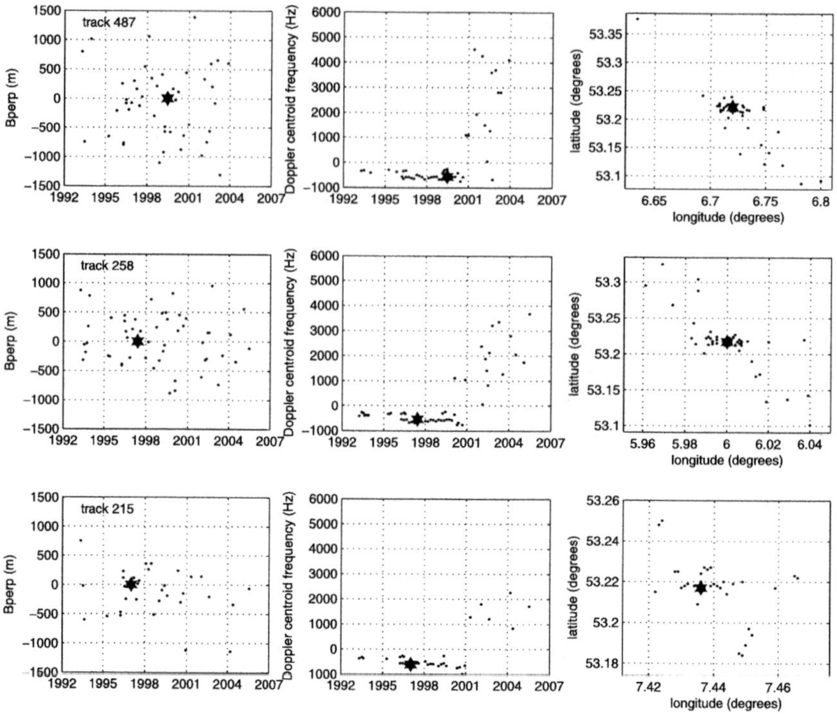

Fig. 6.3 Acquisition geometry of the ascending ERS tracks: temporal baseline versus perpendicular baseline (left), temporal baseline versus Doppler centroid frequency (middle) and geographic location of the center of the scenes (right)

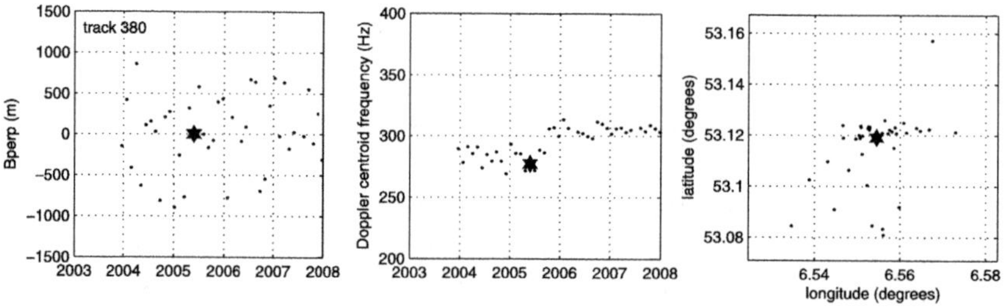

Fig. 6.4 Acquisition geometry of Envisat track 380: temporal baseline versus perpendicular baseline (left), temporal baseline versus Doppler centroid frequency (middle) and geographic location of the center of the scenes (right)

can be determined from (3.11):

$$\Delta D_{ij} = -\frac{\lambda}{4\pi}\frac{2\pi}{v}(f^m_{\text{dc},i} - f^s_{\text{dc},i})\xi^m_{ij}, \tag{6.1}$$

which is 4 mm if an oversampling factor of 2 has been applied.

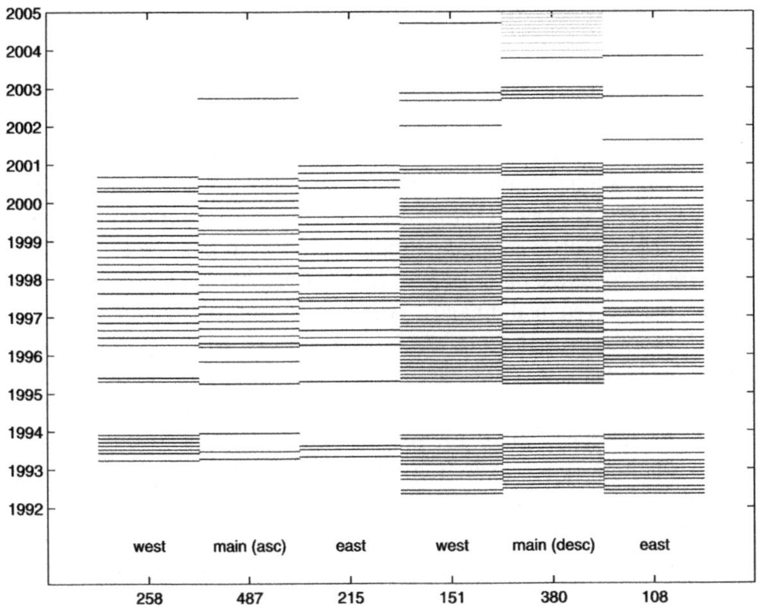

Fig. 6.5 Selected images of the six ERS tracks and the continuity with Envisat (in grey). All ERS-2 images with a Doppler frequency deviation >500 Hz from the master image have been discarded. Due to conflicts with commercial parties, only from Envisat track 380 regular acquisitions were taken in image mode

The temporal distribution of the selected ERS and Envisat images is depicted in Fig. 6.5. The stacks of images per track have been processed throughout the years 2004–2006, and not all ERS-2 acquisitions have been considered from 2004 onwards. For the main tracks 487 and 380, all acquisitions up to the end of 2003 (start of Envisat acquisitions) have been considered in the selections. For all other tracks (151, 108, 258, 215), SAR acquisitions up to medio 2005 have been part of the image selection procedure. After February 2000, when ERS-2 lost its three gyro mode, the SAR images have been acquired with highly varying Doppler frequencies. Consequently, for all three ascending tracks only one SAR acquisition could be used after 2001. The situation for the descending tracks is slightly better, although the average acquisition rate does not exceed one per year. Fortunately, an ERS-2 image with a reasonable Doppler (538 Hz) was acquired on the 21st of December 2003. This is the start date of the Envisat time series: both ERS-2 and Envisat data have been acquired with half an hour time difference. Hence, the Groningen study has extra potential to show the continuity between ERS-2 and Envisat.

Figure 6.5 shows only one Envisat track, 380. Only from this track, sufficient SAR images for a PSI analysis (>25) have been acquired in image mode. This was caused by conflicts of interest with commercial users, who requested data in the vicinity of the area of interest in Wide Swath ScanSAR mode.

6.1.2 Generation of Interferograms

The interferometric processing of the image stacks that are listed in Table 6.1 has been performed according to the procedure that has been described in Sect. 3.1. This section addresses the specific implementation in the Groningen area.

Due to computational limitations in both the interferometric processing of oversampled images and the DePSI estimation procedure, the coverage of each track has been subdivided in four overlapping crops of $\sim$50 $\times$ 50 km, see Fig. 6.6. Only the common ground coverage of all images from a track has been processed, to ensure that each PS is observed in all images. The common area of the four crops has an extent of 5$\times$5 kilometers, and is preferably located in an urbanized area that contains potential PS candidates.

Interferometric processing has been performed up to and including the subtraction of the reference phase. It has already been pointed out in Sect. 3.1.2 that the distribution of coregistration windows and the precision of the coregistration polynomial is essential for precise phase observations and the detection of PS candidates. Since the height differences in the area of interest are less then 30 meters (Fig. 6.7), coregistration has been performed by a 2nd degree polynomial. For all selected interferometric combinations in each track, it has

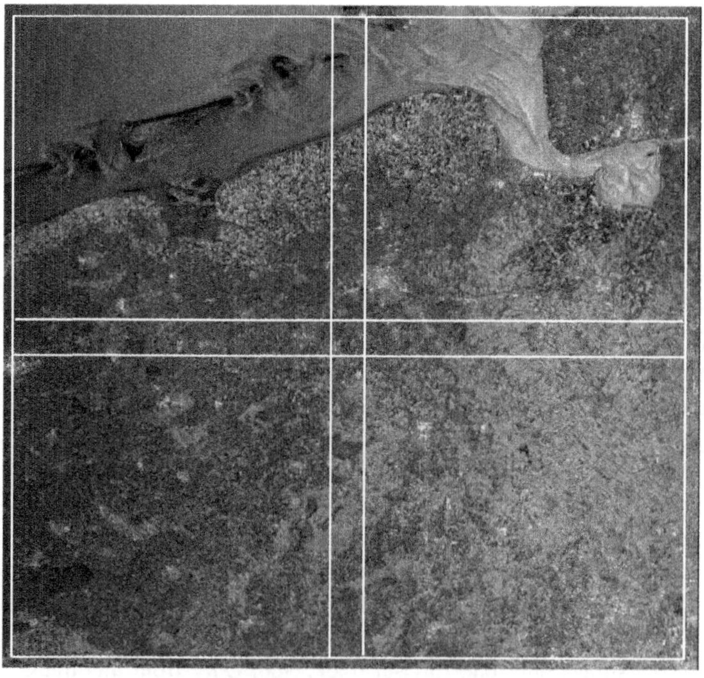

Fig. 6.6 Subdivision of a scene into four overlapping crops. Due to computational limitations, the interferometric processing and the PSI estimation have been performed separately in four overlapping crops. The SAR data has been provided by ESA

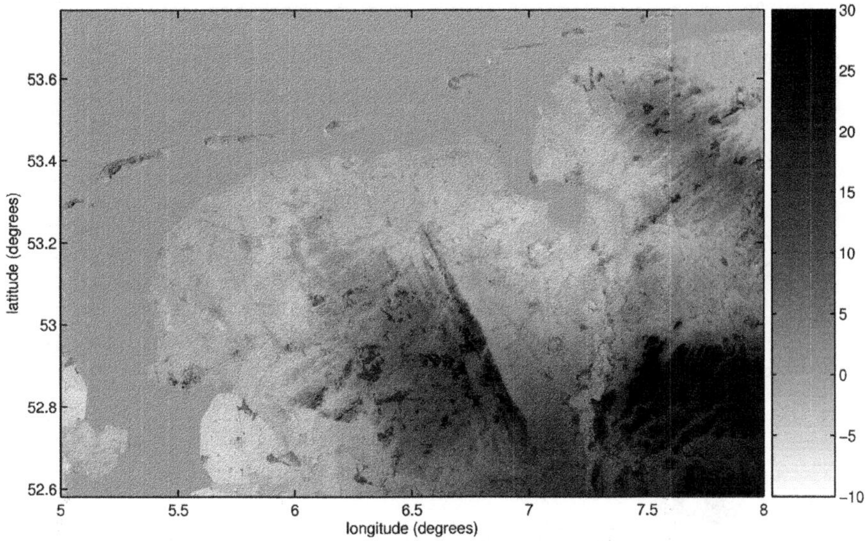

Fig. 6.7 SRTM heights in the northeastern part of the Netherlands (NAVD88 orthometric heights (m)). Data source: SRTM, courtesy NASA/JPL-Caltech

been verified if the coregistration residuals were less than 0.1 pixel and if the accepted coregistration windows were evenly distributed over the area of interest.

The coregistration polynomials of the quarter scenes have been evaluated in the common overlap. The standard deviation of the range and azimuth locations that are computed by the four coregistration polynomials are depicted in Fig. 6.8. The coregistration differences of 92% of the scenes are below 0.1 pixel in the resolution of the original image. The outliers in azimuth direction correspond with scenes that have been acquired after 2000. The correlation with Doppler centroid frequency is not evident: there are examples of images after 2000 with a small Doppler centroid frequency that have location differences larger than 0.1 pixel. This may have an impact on the PS candidate selection and the PSI parameter estimation, see Sect. 6.3.1.

Since the height differences in the area of interest are limited (<30 m, see Fig. 6.7) and topographic heights will be estimated anyway in the PSI estimation procedure, topography has not been subtracted in the interferometric processing of the ERS images. Initially, the subtraction of the topographic phase contribution was omitted for Envisat as well. However, a spatial trend appeared to be present in the Envisat phase observations, which will be discussed in Sect. 6.3.2. It was chosen to estimate and remove this trend prior to PSI estimation, and hence all other possible phase contributions were subtracted to avoid any bias in the spatial trend estimation. Therefore, the topographic phase contributions have been subtracted in the interferometric processing of the Envisat scenes. An external DEM from the Shuttle Radar Topography Mission (SRTM, 2008) was utilized to compute these topographic phase contributions.

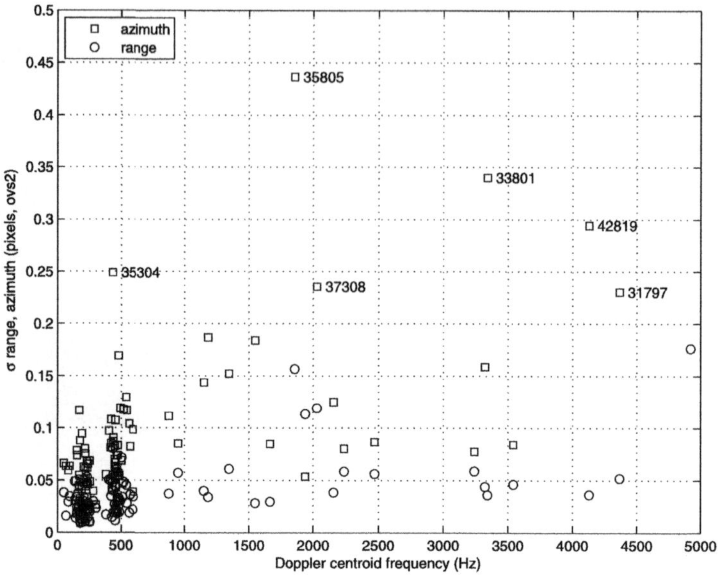

Fig. 6.8 Standard deviation range and azimuth locations (oversampling factor 2) in the common overlap of the four crops, that are independently coregistered. Orbit numbers have been added to the images for which the standard deviation of the range and azimuth locations exceeds 0.1 pixel in original resolution

6.1.3 DePSI

The concepts of the Delft PSI estimation have been introduced in Sect. 3.4. An overview of the implementation in the Groningen study is sketched in Fig. 6.9. PSI estimation has been performed separately for each of the four crops. The reference PS has been chosen in the common overlap.

6.1.3.1 PS Candidate Selection

PS candidates have been selected based on the normalized amplitude dispersion obtained by threshold tuning using pseudo-calibration, see Sect. 3.2. The amplitude threshold for the first order network was set to 0.25, corresponding with a phase standard deviation of 2 mm.

From P PS phase observations, only $P - 1$ independent spatial differences can be formed. However, contrary to traditional geodetic techniques, it is not known beforehand whether a PS candidate is indeed a reliable measurement point or not. Therefore, a (redundant) first order network is constructed to be able to identify falsely selected PS candidates (type II errors). The network construction based on the Delaunay triangulation has been utilized in the Groningen study.

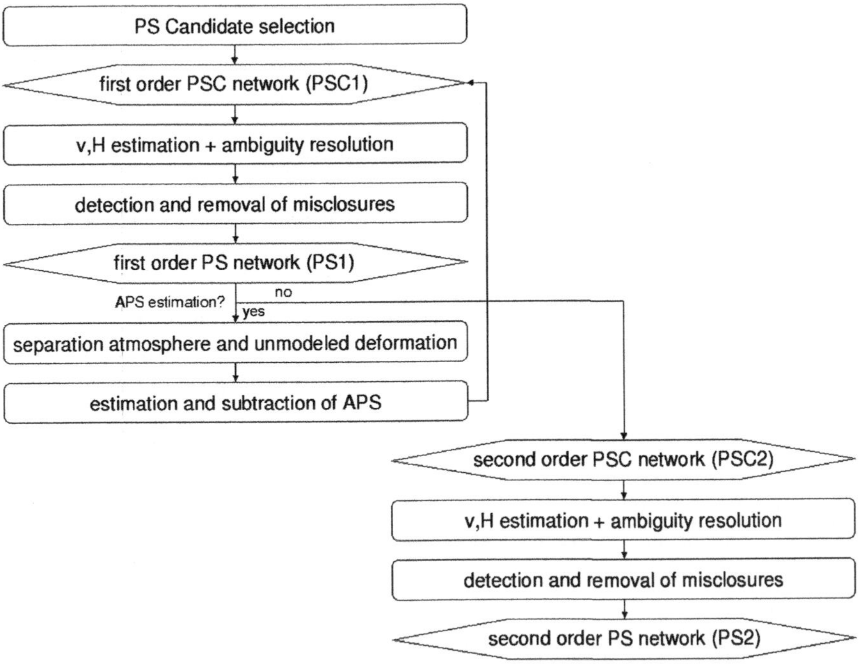

Fig. 6.9 Schematic overview of the DePSI estimation procedure for the Groningen study

6.1.3.2 Parameter Estimation in the First Order Candidate Network

Parameter estimation in the initial network is performed per arc, according to (3.11) and (3.12). Several options for deformation parameterization are available: linear or periodic in time, over the entire observation period or for a selected time window only. The latter models a breakpoint in the displacement time series. It can for example be used to model the delayed onset of subsidence in the monitoring period. If no a-priori information is known about the deformation signal, a constant velocity is the strongest model in terms of redundancy. Since the majority of the Groningen area is subsiding with a (near) constant rate (see Fig. 2.5), the system of equations for an arc between PS i and PS j reads:

$$E\left\{\begin{bmatrix}\underline{\varphi}_{ij}^{k=1}\\ \vdots \\ \underline{\varphi}_{ij}^{k=K}\\ \underline{v}\\ \underline{H}\end{bmatrix}\right\} = \begin{bmatrix} -\frac{4\pi}{\lambda}T^k & -\frac{4\pi}{\lambda}\frac{B_i^{\perp}}{R_i^m \sin\theta_i^m} & -2\pi \\ 1 & 0 & 0 \\ 0 & 1 & 0 \end{bmatrix}\begin{bmatrix} v\\ H\\ a\end{bmatrix}, \tag{6.2}$$

with the corresponding variance-covariance matrix:

$$D\left\{\begin{bmatrix} \underline{\varphi}_{ij}^{k} \\ \underline{v} \\ \underline{h} \end{bmatrix}\right\} = \begin{bmatrix} Q_{\varphi} & 0 & 0 \\ 0 & \sigma_{v}^{2} & 0 \\ 0 & 0 & \sigma_{H}^{2} \end{bmatrix}, \tag{6.3}$$

where K is the number of interferograms, v is the relative displacement rate between PS i and j, T^k is the temporal baseline for interferometric combination k, and H is the topographic height difference between PS i and j.

The variance-covariance matrix Q_y is composed of the variance-covariance matrix of the observations and the variance matrix of the pseudo-observations. The variances of the pseudo-observations determine the search space in the ambiguity resolution. For the Groningen study, the standard deviation of the displacement rates and height differences per arc have been set to 20 mm/year and 30 m respectively.

The variance-covariance matrix of the double-difference phase observations Q_{φ} is a superposition of measurement noise and model imperfections due to atmospheric signal and unmodeled deformation (Hanssen, 2001; van Leijen et al., 2006a). Since all double-differences are mutually correlated, this is a full matrix. The temporal correlation length of unmodeled deformation has been set to one year, which corresponds with the length of the moving average window that has been used to separate atmosphere and unmodeled deformation. The variance-covariance matrix is not well known a-priori, and can only be updated after ambiguity resolution due to the lack of redundancy in (6.2).

To reduce computation time, integer bootstrapping has been applied to resolve the integer ambiguities. The success rates depend on the a-priori stochastic model. The variances of the pseudo-observations are usually set relatively large to guarantee that the solution is within the search space. Because there is no redundancy in the system of equations, the a-priori stochastic model cannot be improved by variance component estimation. Hence, an objective evaluation of the success rates is not possible.

6.1.3.3 Unwrapping Test Procedure and Redundant Parameter Estimation

A spatial unwrapping test procedure is performed to detect and remove arcs that exhibit ambiguity misclosures (van Leijen et al., 2006b). The first order candidate network is considered as a leveling type of network with the ambiguity estimates as the observations. Only the ambiguities of arcs above a temporal coherence threshold (see (3.7)) of 0.6 are selected in the Groningen study. ‘Dead end’ arcs are removed from the set, since they cannot be tested and are therefore unreliable. After the geodetic testing procedure thats detects and removes ambiguity outliers, the accepted ambiguity estimates are considered deterministic and a redundant parameter estimation is performed. All double-difference phase observations are now unwrapped with respect to one reference point. Equation (6.2) can be simplified to the following redundant system:

$$E\begin{bmatrix} \underline{\varphi}_{ij}^{k=1} \\ \vdots \\ \underline{\varphi}_{ij}^{k=K} \end{bmatrix} = \begin{bmatrix} -\frac{4\pi}{\lambda}T^k & -\frac{4\pi}{\lambda}\frac{B_i^{\perp}}{R_i^m \sin\theta_i^m} \end{bmatrix}\begin{bmatrix} v \\ H \end{bmatrix}; \qquad D\left[\underline{\varphi}_{ij}^k\right] = Q_{\varphi}, \tag{6.4}$$

where $\underline{\varphi}_{ij}^k$ are now the unwrapped double-difference phase observations. The redundancy equals $(K-2)$ per arc, and VCE can be applied to obtain a more realistic variance-covariance matrix. For a single arc, measurement noise cannot be distinguished from atmospheric signal. Hence, only one variance factor per arc is estimated from (6.4), that represents both measurement noise and residual atmospheric disturbances.

6.1.3.4 Separation of Atmospheric Signal and Unmodeled Deformation

The residuals of the parameter estimation of the first order network are separated into atmospheric signal and unmodeled deformation. This separation is based on the assumption that atmospheric signal is not correlated in time whereas unmodeled deformation is. First, the atmospheric phase screen of the master is estimated. It is present in each interferogram and equal to the average of the residuals in time. After subtraction of the master APS, the phase residuals are divided into atmospheric signal and unmodeled deformation, as explained in Sect. 4.3.2.

If the assumption is valid that the deformation signal in the area of interest has a constant displacement rate over the entire time span, high frequency variations in time are addressed to atmospheric signal. However, in practice we are more conservative, since deviations from the linear displacement model should not leak into the atmospheric signal. Deformation residuals can contain valuable information, such as settling effects, subsidence delay after start of gas extraction, uplift due to underground gas storage, and seasonal effects. Hence, only residuals with a temporal correlation length smaller than one year are attributed to atmospheric signal in the Groningen study.

6.1.3.5 Interpolation of Atmospheric Residuals by Means of Kriging

By means of Kriging, the residuals addressed to the atmospheric signal are spatially interpolated. As explained in Sect. 4.2.4, the atmospheric regimes to be considered are regime I that covers large scale variations, and regime II that covers scales from resolution level to the thickness of the turbulent layer. For rural areas, it has to be taken into consideration that regime II, that covers scales lower than 2 kilometers, may be undersampled. In the Groningen crops, the number of accepted arcs in the initial network varies between 2000 and 5000, over an area of $\sim$2500 km^2. On average that is 1–2 arc/km^2, but in practice the density in urban areas will be higher and in rural areas it will be lower. The Kriging procedure cannot reconstruct under-sampled signals and as a result, atmospheric disturbances in the lower scales will be smoothed, if present. Consequently, the precision of the PSI parameter estimates will be lower in rural areas, see Sect. 6.3.1.

6.1.3.6 PS Densification: The Second Order Network

After subtraction of the atmospheric phase contribution estimated from the first order network, a further densification step is performed by the analysis of the phase history of PS candidates with a higher normalized amplitude dispersion. Parameter estimation is performed for each PS candidate with respect to the three closest accepted PS from the first order network. Subsequently, these estimates are checked on their consistency. The PS candidate under investigation is accepted if the estimates of at least two out of three connecting arcs are in agreement.

6.2 ERS and Envisat PSI Results

PSI estimation has been applied to six overlapping ERS tracks and one Envisat track covering the subsiding area above the Groningen gas field (see Table 6.1). Here we discuss the PSI deformation estimates that have been obtained from the interferometric processing and PSI estimation, as described in Sects. 6.1.2 and 6.1.3.

6.2.1 ERS Deformation Estimates

Figure 6.10 shows the obtained deformation estimates for the ascending and descending main tracks over the Groningen subsidence area. From the double-difference phase observations, PS velocities have been estimated in the period 1992–2003. The PS velocity estimates clearly depict the subsiding areas due to gas extraction. Although the PS density is relatively low in rural areas, the PSI technique is certainly able to estimate a coherent pattern of earth surface movements. The relative displacement rates are in agreement with the expected velocities for this type of subsidence (-7 up to 2 mm/year). For comparison, the leveling heights have been converted to displacement rates as well. The PSI and leveling estimates are visually in agreement. A quantitative analysis will be performed in Chap. 7.

The average PS density in the area of interest is $\sim$40 PS/km^2, but these PS are not evenly distributed. The PS density is varying from up to 0–10 PS/km^2 in the rural areas to over 100 PS/km^2 in the urbanized areas, see Fig. 6.11. The PS distribution follows the urbanized areas. Figure 6.12 shows that PS targets in rural areas coincide with buildings and structures. This implies that when these man-made features are absent, the PS density drops to 0 PS/km^2. In the Groningen subsidence

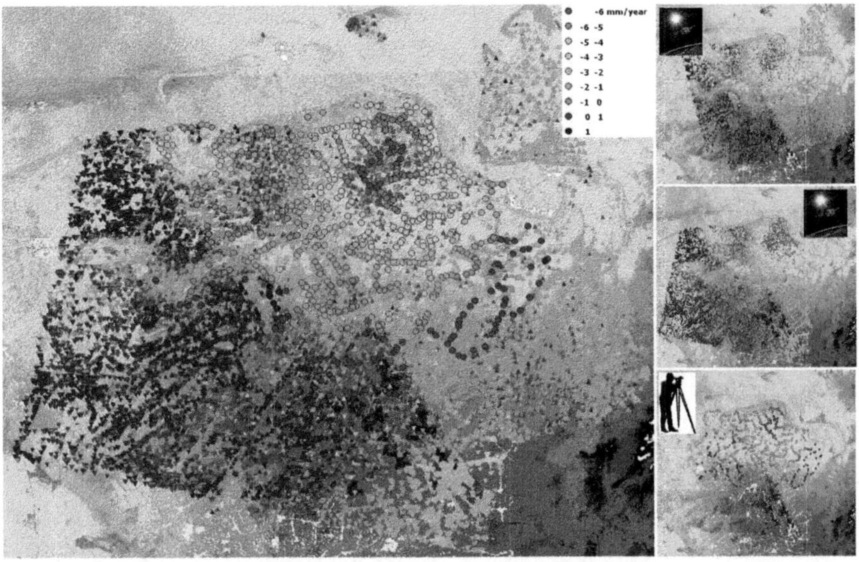

Fig. 6.10 Persistent Scatterer (triangles) and leveling benchmark (circles) displacement rates (mm/year) in the period 1992–2003 in the northeastern part of the Netherlands. The PSI displacement rates stem from the main tracks 380 (descending) and 487 (ascending). Background data source: SRTM

Table 6.2 PS density over the Groningen subsidence area that is characterized by agricultural fields with scattered farms and small villages

Track	0 PS/km^2	1–5	5–25	25–100	>100
380 (ERS)	20%	19%	28%	27%	6%
487 (ERS)	18%	13%	32%	29%	9%

area, this is the situation in ~20% of the total area for a single satellite track, see Table 6.2. By means of the combination of multiple tracks, the PS density can be improved, see Sect. 6.4. Other methodologies that may lead to a higher PS density are supervised PS selection (Humme, 2007), and the application of adaptive deformation models (van Leijen and Hanssen, 2007). Moreover, the utilization of new sensors with a smaller wavelength and higher resolution, such as TerraSAR-X, may lead to a higher PS density. Still, an average PS density of ~40 PS/km^2 in a dominantly rural area is high compared to the density of leveling benchmarks (1–2 per km^2). Since these leveling benchmarks are distributed along the existing infrastructure, the leveling technique will also leave areas with solely agricultural fields uncovered.

Figure 6.13 shows the temporal sampling of InSAR. Although deformation is modeled as a linear, constant displacement rate (velocity), the displacement estimates show the actual behavior of the target. The a-priori modeling of deformation as a velocity is only used to correctly unwrap the PS phase observations,

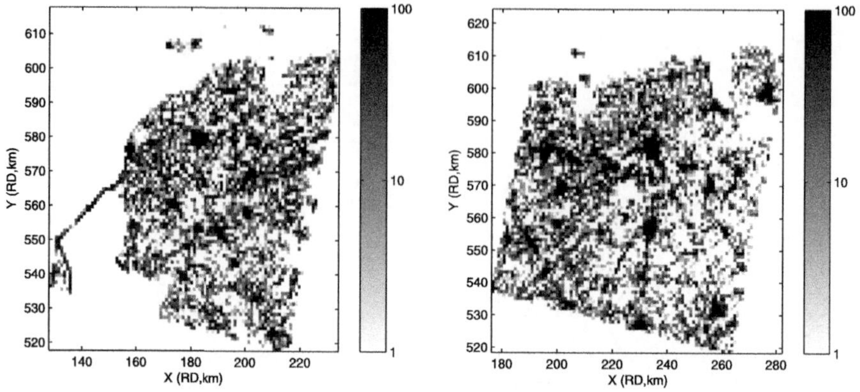

Fig. 6.11 Typical PS density distributions in the northern part of the Netherlands: track 151 (left) and track 380 (right). The PS density in the rural areas is 0–10 PS/km^2, whereas the PS density in the urbanized areas is >100 PS/km^2

Fig. 6.12 PS targets in a rural area from four overlapping tracks (ascending, descending, and adjacent). Almost every building serves as a Persistent Scatterer. Map: © 2007 Google Earth, Image © 2007 Aerodata International Surveys, © 2007 Europa Technologies

see Fig. 3.11. Changes in displacement rate are still captured in the actual unwrapped displacements. However, the likelihood of correct unwrapping is maximized when using adaptive deformation models (van Leijen and Hanssen, 2007). Here, advanced deformation models are validated as alternative hypotheses. The most likely deformation model is subsequently selected. Adaptive deformation models are not required a-priori in the Groningen subsidence area. By means of

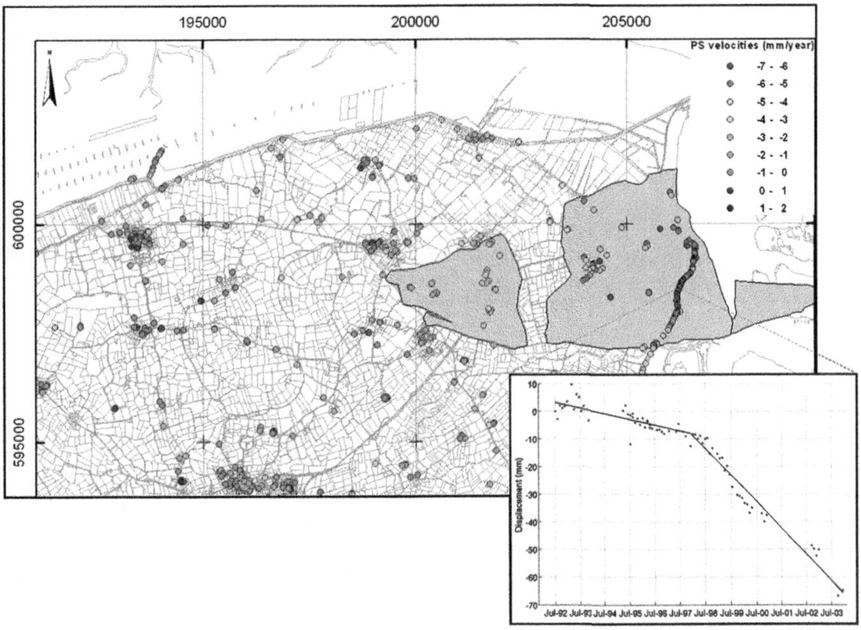

Fig. 6.13 The Anjum gas field (in green) that has been taken into production medio 1997; PS displacement rates in the period 1992–2003; displacement time series and a potential alternative hypothesis for a single PS target. This PS target shows a trend change when subsidence due to gas extraction starts. Hence, the depicted average displacement rates underestimate the actual subsidence due to gas extraction. A possible explanation of the PS displacement rate before 1997 is shallow compaction. Source topographic background: © Kadaster en de Openbare Registers, Apeldoorn. Overlay created in ArcGIS®

residue analysis of multiple PS displacement time series, deviations from the constant velocity model can be detected (Ketelaar et al., 2006). It is recommendable to apply an alternative hypothesis in areas where all PS exhibit the same deviation from the constant velocity model. Such an area is the Anjum area (see Fig. A.1), where gas production has started medio 1997. The change in subsidence rate in Fig. 6.13 corresponds with the onset of gas extraction. Due to the high temporal sampling of PSI displacements, the delay between the start of gas extraction and start of subsidence can be determined, which lies between several months and one year. This is new valuable information that can be detected from PSI, which was not possible with periodic levelings each 2–5 years. The application of PSI for monitoring reservoir behavior will be further addressed in Chap. 8.

Not only the deformation estimates of the ERS main tracks capture deformation signal that coincides with the regions that are subsiding due to gas extraction. Figure 6.14 shows the PS velocity estimates in the period 1992–2005 for all six tracks covering the Groningen subsidence area: the main tracks plus the four adjacent tracks. They all independently detect the areas that are affected by ground

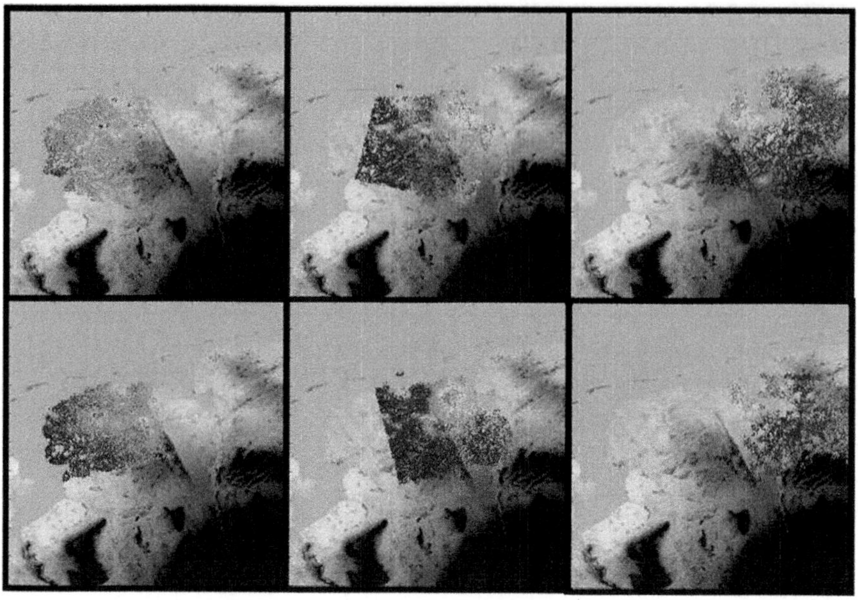

Fig. 6.14 PSI velocity estimates (mm/year) of all six ERS tracks. All tracks depict the subsiding areas. Although they refer to a different reference PS, the relative velocity estimates are similar. Background data source: SRTM

movements. Although each track has a different reference PS, the relative displacement rates are visually in agreement. A quantification of the misclosures between the tracks is performed in the multi-track datum connection procedure, see Sect. 6.4.

6.2.2 Envisat Deformation Estimates

As was pointed out in Sect. 6.1.1, only one Envisat track (380) has been monitored sufficiently in image mode to perform PSI. Its first image has been acquired in December 2003. With only half an hour time difference, an ERS-2 image was acquired as well, with a Doppler centroid frequency of 538 Hz. Hence, the continuity of satellite deformation monitoring is secured over the area of interest. Contrary to the ERS-2, the Envisat phase observations have to be corrected for a spatial trend, which will be explained in Sect. 6.3.2.

Figure 6.15 shows the Envisat PS velocity estimates of track 380 in the period 2003–2007. Again, the subsiding areas are clearly depicted. The relative velocities are of the same magnitude compared to ERS in the period 1992–2003: up to −7 mm/year. The ERS-2 and Envisat displacement estimates can be connected in the joint estimation of the deformation signal. Figure 6.16 shows the displacement time series of two nearby PS that have been merged in the estimation of the displacement rate. Besides connection of deformation estimates in the parameter of interest space, it may be possible to identify common PS targets that are monitored by both

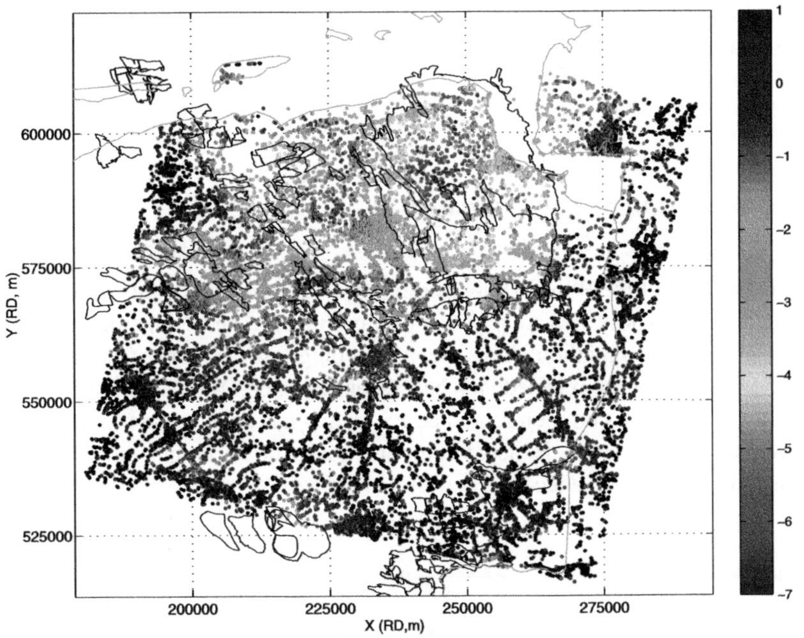

Fig. 6.15 Envisat PS velocity estimates (mm/year) in the period 2003–2007, after detrending and outlier removal. Coordinates are in the Dutch RD system. The outline of the gas fields are depicted in black; the country borders are indicated with grey lines

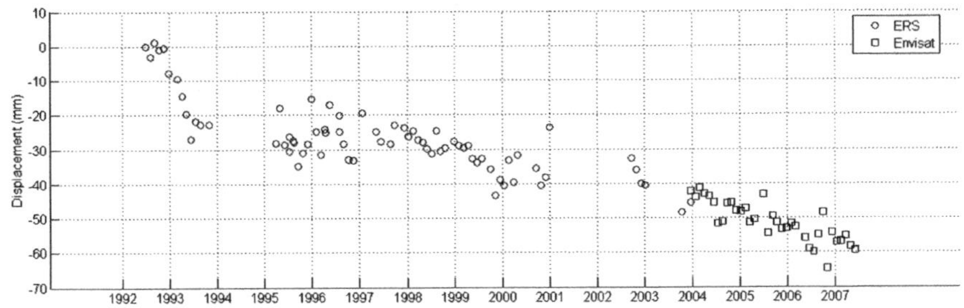

Fig. 6.16 Time series that consists of both ERS (circles) and Envisat (squares) displacement estimates. The displacement estimates have been connected in the joint estimation of a linear displacement rate

sensors, which is recommended to further investigate in the future. A comparison of the Envisat PS displacements and predicted displacements from historic leveling campaigns is performed in Chap. 7.

6.3 Quality Control

This section addresses the precision of the PSI estimates of the tracks that cover the Groningen area. Furthermore, it will investigate the spatial trend in the Envisat phase observations. The quantification of the precision of PSI deformation estimates in this chapter is based on the results of the C-band missions ERS-1, ERS-2 and Envisat.

6.3.1 Precision of PSI Estimates

The estimation of the PS velocities and relative topographic heights have been performed per arc. These arcs are part of a network that has been constructed by means of a Delaunay triangulation. After the unwrapping test procedure (see Sect. 6.1.3), all observations of the rejected arcs have been removed. The unwrapped phase observations refer to a common reference PS, which has been chosen in the common overlap of the four crops. Hence, within one track, the PS height and velocity estimates refer to one reference PS.

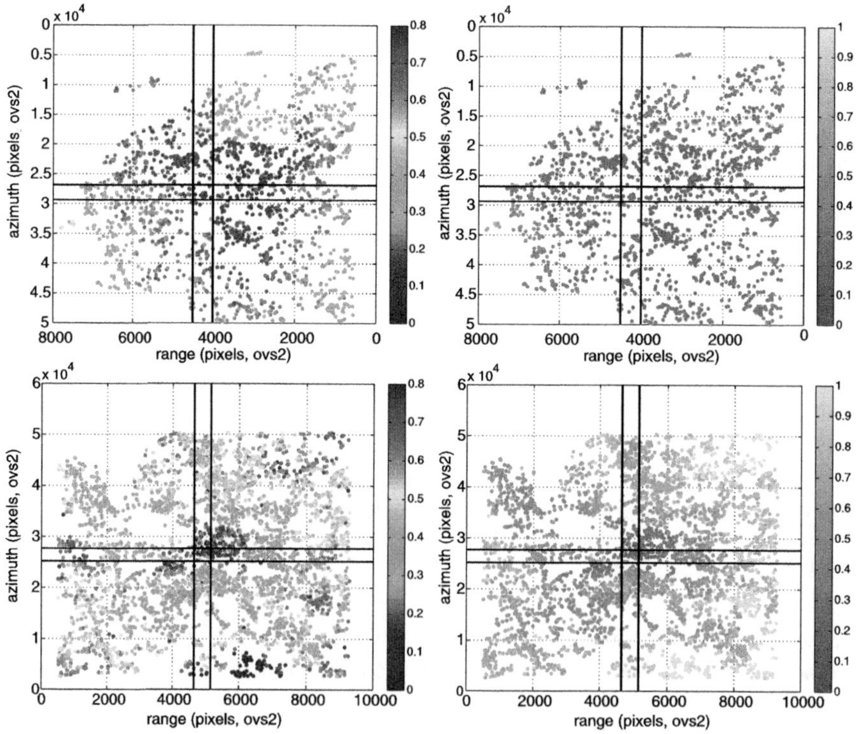

Fig. 6.17 Top: precision of PSI estimates for track 151 (74 interferograms): standard deviation of PS velocities (left) and standard deviation of topographic heights (right). Bottom: precision of PSI estimates for track 215 (24 interferograms): standard deviation of PS velocities (left) and standard deviation of topographic heights (right). The reference PS is located in the center of the scene for both tracks. The black lines indicate the overlap of the four crops

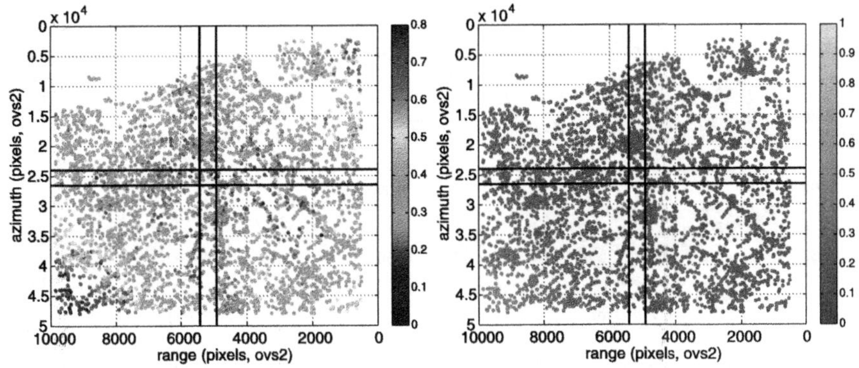

Fig. 6.18 Precision of the Envisat PSI estimates for track 380 (40 interferograms): standard deviation of PS velocities (left) and standard deviation of topographic heights (right). The reference PS is located in the center of the scene. The black lines indicate the overlap of the four crops

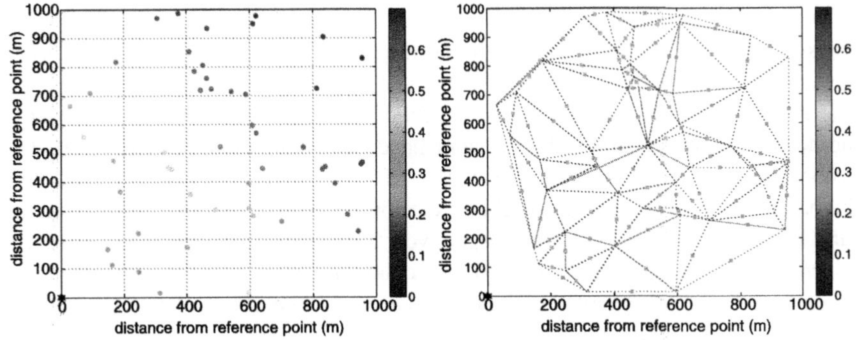

Fig. 6.19 Simulation of precision PSI velocity estimates (mm/year). If only the variances are considered, the precision seems to decrease further away from the reference PS (left). However, the relative precision of nearby PS is similar irrespective of the location (right)

Figure 6.17 shows the standard deviation of the velocity and height estimates after variance component estimation for the tracks with the largest and the smallest number of interferograms. The standard deviation of the velocity estimates scales with the distance to the reference PS. For the smallest Groningen stack consisting of 24 interferograms, the standard deviation is $\sim$0.1 mm/year per $\sqrt{\text{km}}$; for the largest stack of 74 interferograms it is $\sim$0.04 mm/year per $\sqrt{\text{km}}$.

Interpretation of the standard deviation as an absolute precision measure gives a distorted impression. A PS further away from the reference PS is not of lower quality. By taking linear combinations of the parameter estimates of neighboring PS and applying the propagation law of variances and covariances, it can be shown that the relative precision is location independent. For example, Fig. 6.19 illustrates the precision within a simulated PS network. The relative precision in the network is similar for arcs of equal length.

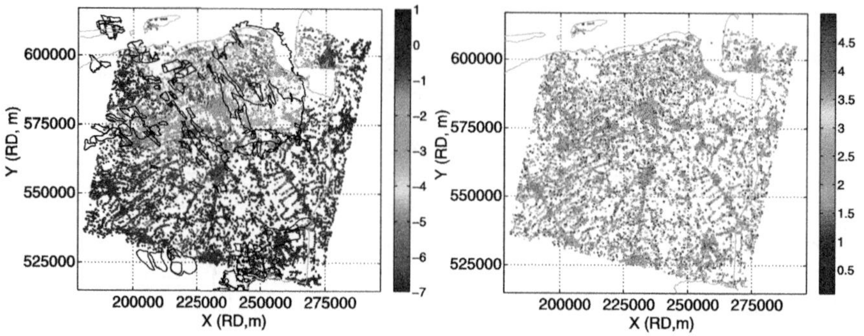

Fig. 6.20 PSI velocity estimates for Envisat track 380 (left) and standard deviation of PS displacements with respect to the closest PS in the first order network (right). The relative displacement precision is independent of the choice of the reference PS, but decreases with increasing distance between the PS (in rural areas). The displacement standard deviation in urban areas is ≤ 3 mm; in rural areas it is 3–7 mm

Due to the PS density, the sampling of the atmospheric signal is higher in urban areas compared to rural areas. Moreover, the arc lengths in rural areas are larger. Here, we investigate if this results in systematically less precise parameter estimates. Therefore, the standard deviations of the PS displacement residuals with respect to the nearest PS in the first order network have been computed, see Fig. 6.20. The displacement precision of the majority of the PS is ~ 3 mm (1-sigma). As expected, the parameter estimates have a higher precision in the urban areas compared to the rural areas (≤ 3 mm and 3–7 mm respectively). However, although the precision is lower, the rural areas are still covered with coherent velocity estimates.

Besides the precision of the displacement errors, the effect of potential model errors, such as unwrapping errors, is investigated. Since the PSI estimation of each track has been performed for four overlapping crops independently, the overlapping areas can be utilized for a reliability check. The differences between the velocity, height and displacement estimates in the overlap should be within their precision limits. Figure 6.21 shows the results for the largest stack (track 151, 74 interferograms) and the smallest stack (track 215, 24 interferograms).

The standard deviation of the PSI velocity estimates in the overlap of the four crops (Fig. 6.21) are equal to the standard deviations that are depicted in Fig. 6.17. Figure 6.22 gives a further insight in the standard deviation of the displacements. It can be seen that the precision of the displacement estimates from the small stack is lower than the precision of the large stack. Furthermore, displacement outliers are visible in the period after 2000, that is only covered by track 151. In track 215, no acquisitions were available with a low Doppler deviation after 2000, see Fig. 6.5. As a result, the temporal sampling is high over the entire monitoring period. Track 151 contains five acquisitions in the years 2001–2005. The low temporal sampling in this period decreases ambiguity resolution success rates. Since all observations of a single arc are removed when an ambiguity in a single epoch is rejected, this leads to a lower PS density in the first order network. Consequently, the APS estimates may be affected as well. Although track 151 contains significantly more acquisitions

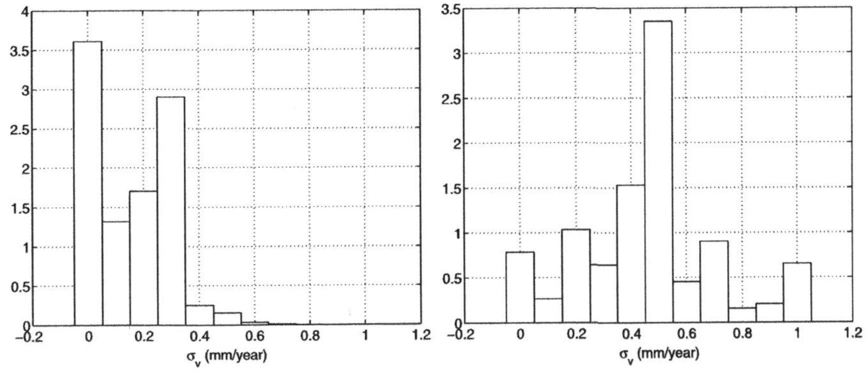

Fig. 6.21 Histograms of the standard deviations of the PS velocity estimates in the overlap of the four crops for track 151, that spans the period 1992–2005 (left), and for track 215, that spans the period 1993–2000 (right)

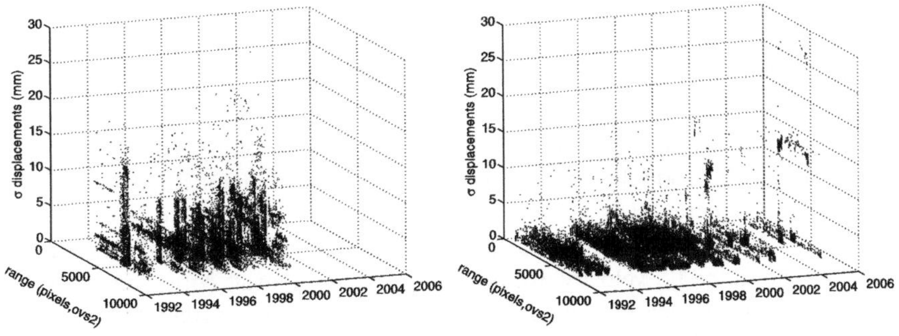

Fig. 6.22 Precision of displacement estimates (mm) for track 215 (24 interferograms) and track 151 (74 interferograms). The precision of the displacement estimates from track 215 is on average lower. Furthermore, it can be seen that the acquisitions after 2000 (track 151) exhibit large displacement residuals

than track 215, the precision of the displacement rates can be slightly deteriorated by including the sparse acquisitions after 2000. An option would be to remove all acquisitions after 2000, but at the same time the sparse observations in this period can contain valuable information for the continuation of subsidence monitoring. Hence, they have been included in the PSI estimation for the Groningen area.

6.3.2 Unmodeled Residual Components

In Sect. 4.2, the effect of shortcomings in the functional model has been addressed: the influence of the sub-pixel position, sidelobe observations, and orbital inaccuracies on the PSI parameter estimates have been investigated. It has been shown that orbital inaccuracies can lead to a spatial trend in the PS velocity estimates from near to far range. The velocity difference in radial and across-track direction can be up to respectively ~1 and ~0.5 mm/year.

Unmodeled residual components are further investigated in this section, since a spatial trend appeared to be present in the Envisat PSI velocity estimates of ~15 mm/year over a distance of 100 kilometers. Since this spatial trend is unlikely to be caused by any of the model errors that are discussed in Sect. 4.2, the origin of this spatial trend is now further investigated.

6.3.2.1 Envisat PSI Estimates

The PSI processing of the Envisat descending main track has been performed in a similar way as for ERS. The master scene has been subdivided in four overlapping crops that each cover an area of ~50×50 km. PS candidates have been detected based on normalized amplitude dispersion using pseudo-calibration. Subsequently, a first order network was formed, followed by unwrapping and parameter estimation using integer bootstrapping for ambiguity resolution. After APS estimation and subtraction, the PS network was further densified into a second order network.

The results of all four crops demonstrate that Envisat is able to detect coherent patterns of local subsidence, but that an artificial trend appears to be superposed on the PS velocity estimates, see Fig. 6.23. This trend is in the order of 15 mm/year over a distance of 100 kilometers. Since it is not present in the ERS estimates, it is unlikely that it represents real deformation signal. The spatial trend in the Envisat velocity estimates appears to be a function of range and azimuth direction. Figure 6.23 shows the PS velocity estimates after trend removal. After correction, the subsiding areas are clearly visible, and the velocity rates are in agreement with those estimated from ERS. Since the maximum subsidence rate in the Groningen subsidence bowl is −7 mm/year, the potential causes for the systematic spatial trend need to be identified.

Systematic effects in the parameter estimates can be caused by inaccurate APS estimates. However, the PS residuals in the first order network and the APS are in agreement. Moreover, multiple iterations for atmosphere estimation decrease the residuals (from several mm to sub-mm level), but hardly affect the velocity estimates

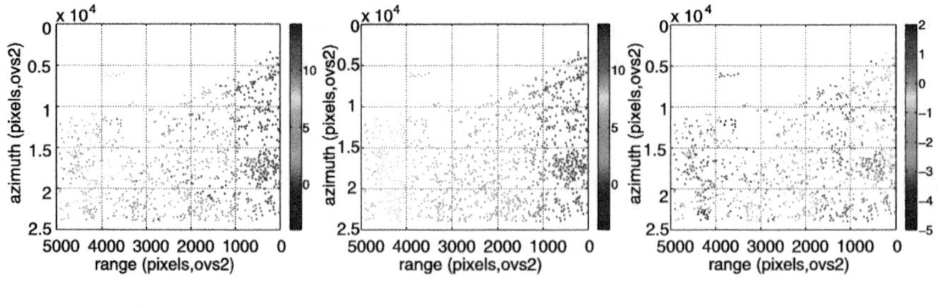

Fig. 6.23 Original PS velocity estimates in mm/year (left); estimated spatial trend as a function of range and azimuth coordinates (middle); PS velocity estimates after subtraction of the estimated spatial trend (right)

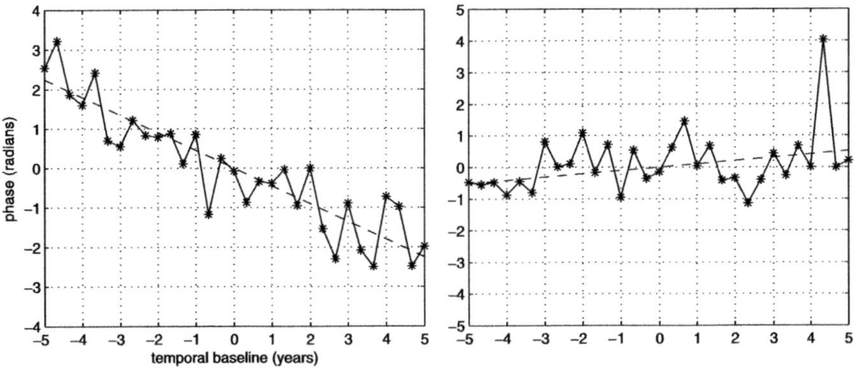

Fig. 6.24 Simulated phase observations between two PS. These phase observations are assumed to be representative for each combination of two PS at similar locations in the radar coordinate system. The PS velocity estimates are affected if the phase observations exhibit a systematic development in time (left) and to a lesser degree if a spatial trend is present in one interferogram that acts as an outlier (right)

(<0.1 mm/year). Hence, it is concluded that the superposed trend on the Envisat PSI estimates is not induced by inaccurate estimation of one or more APS. Furthermore, the velocity estimates are not significantly affected by varying the temporal correlation length of unmodeled deformation. Consequently, the conclusion is drawn that the spatial trend is already present in the velocity estimates per arc in the first order network, prior to APS estimation and the separation of unmodeled deformation. Since the PS velocities are a function of time, this implies that the systematic trend has a component in time. It is likely that the spatial trend exhibits a systematic development in time. The other option is the presence of one or a few interferograms with a spatial trend in the phase observations. However, Fig. 6.24 shows that the presence of such 'outlier' interferograms has less effect on the velocity estimates than a systematic development of the spatial trend in time.

In order to trace the origin of the spatial trend in the Envisat velocity estimates, the Envisat interferometric phase observations are inspected. Figure 6.25 shows two of the interferograms (quarter scenes). The temporal baseline is 35 days, and a spatial trend is visible. Although the majority of the interferograms appear largely decorrelated at first glance due to temporal decorrelation, they contain numerous PS candidates. The phase observations of the PS candidates can be used for the estimation of a spatial trend per interferometric combination, provided that the fringe frequency that causes the trend is sufficiently sampled. In other words, the minimum PS distance should be smaller than half of the spatial fringe frequency. Since the Envisat spatial trends are relatively small (maximal 1–2 fringes in a quarter scene), this condition holds. Hence, the phase observations can easily be unwrapped. After unwrapping, the spatial trend can be estimated as a plane or polynomial as a function of the radar coordinates. Subsequently it can be subtracted from the original phase observations, see Fig. 6.26.

In stead of performing spatial trend removal prior to the PSI estimation procedure, the functional model could be extended with additional unknown parameters

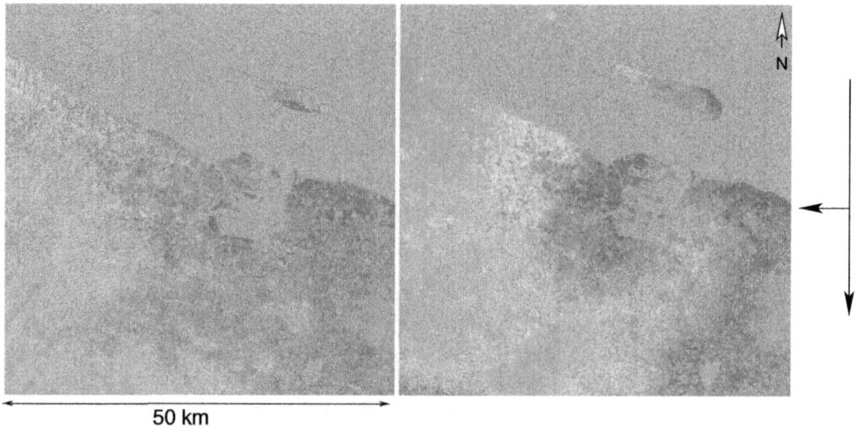

Fig. 6.25 Spatial trends in Envisat interferograms with a temporal baseline of 35 days (quarter scenes). The SAR data has been provided by ESA

that describe the spatial trends. However, PSI parameter estimation is performed per arc between neighboring PS using a non-redundant system of equations. Additional unknown parameters would require additional pseudo-observations. Furthermore, the trend parameter estimates are more precise when using observations over a large spatial range. Therefore, spatial trend removal is currently implemented as an independent step prior to the actual PSI estimation procedure.

Since the spatial trend removal is performed prior to the PSI estimation, the phase observations that are used for the spatial trend estimation contain contributions due to deformation, topography, atmosphere and noise. To minimize the influence of the topographic phase contribution in the estimation of the spatial trend, it is subtracted using an external DEM. The influence of atmosphere and deformation signal can be minimized by the application of a testing procedure that removes spatially correlated outliers in the spatial trend estimation. This works, provided that the majority of the area that is covered in the scene is not affected by deformation. Since this holds for the Groningen study, this strategy has been applied. Another option for the reduction of deformation signal in the spatial trend estimation would be the use of interferometric combinations with minimum temporal baseline.

Summarizing, the PSI procedure for Envisat differs from the ERS procedure in the following way:

- topographic signal is subtracted using an external DEM,
- a spatial trend in range and azimuth direction is estimated and removed based on the phase observations of the PS candidates prior to the actual PSI estimation procedure.

Figure 6.26 shows the results of this procedure. After the correction of the phase observations, the spatial trend in the PS velocity estimates is absent, and the subsidence

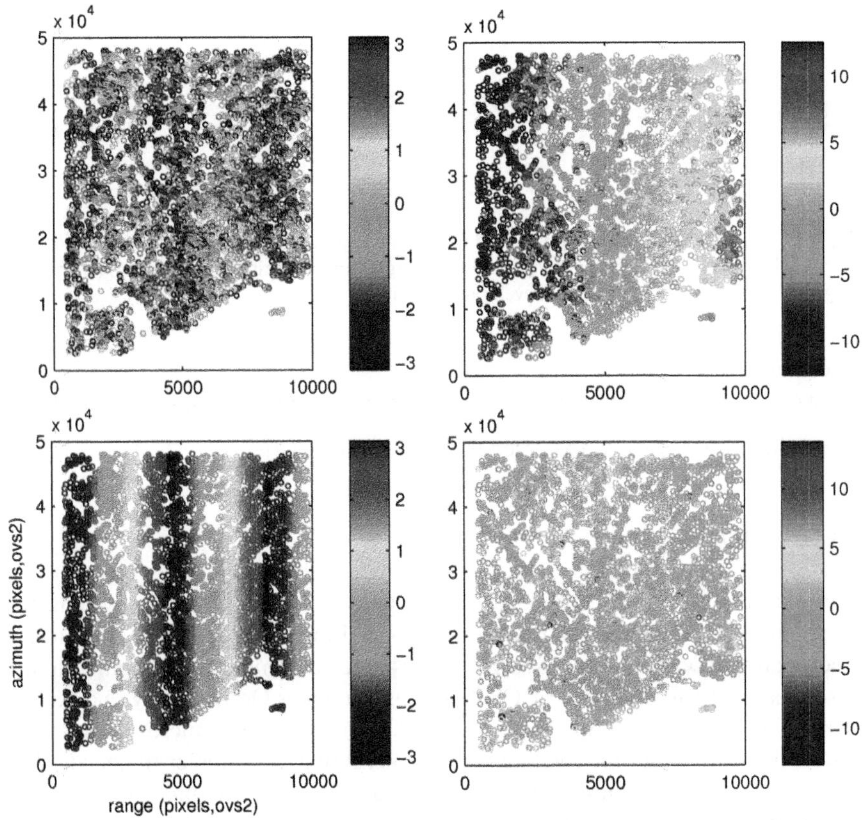

Fig. 6.26 Trend estimation for one interferogram based on the phase observations (in radians) of PS candidates. Top: original wrapped phase observations (left) and unwrapped phase observations (right). Bottom: estimated spatial trend as wrapped phase observations (left) and phase observations that have been corrected for the spatial trend (right)

rates are in the order of those computed from ERS, see Fig. 6.15. The precision of the PS velocity and height estimates are depicted in Fig. 6.18.

Now it has been shown that the spatial trends in the Envisat phase observations can be estimated and removed, the question remains which phenomenon causes the spatial trends. Hence, the temporal development of the spatial trends is investigated. Figure 6.27 shows the range differences over 100 kilometers in range and azimuth direction, computed from the unwrapped phase observations. These range differences can be interpreted as deformation along line of sight. From Fig. 6.27 can be concluded that the spatial trends in range direction exhibit a systematic development in time.

The physical causes for the spatial trends that are present in the Envisat phase observations are subdivided into two groups:

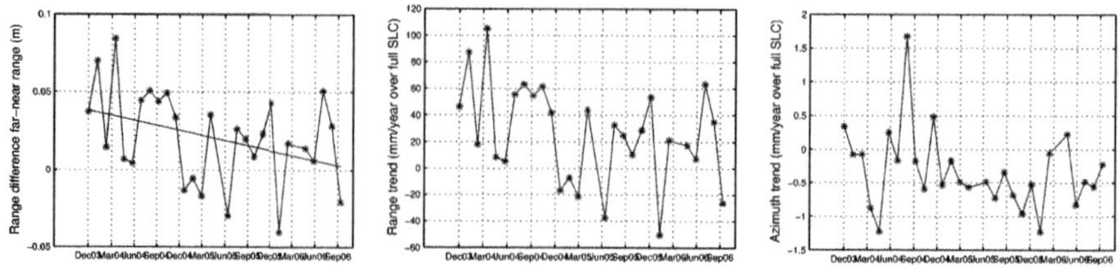

Fig. 6.27 Trend estimates from unwrapped PS candidate phase observations at 100 km distance: (left) total error, (middle) error in range direction, (right) error in azimuth direction. The errors in range direction exhibit a systematic development in time

1. orbital inaccuracies,
2. errors in system parameters (e.g., timing errors and range sampling rate).

They will be successively addressed in this section.

6.3.2.2 Orbits

The precision of the orbits depends on the precision of the tracking system, the precision of the reference station coordinates and the gravity model. The Envisat precise orbits (Doornbos and Scharroo, 2004) are determined using both Satellite Laser Ranging (SLR) and the DORIS system (Doppler Orbitography and Radio-positioning Integrated by Satellite (DORIS, 2008)). DORIS is a microwave tracking system that determines the satellite orbit by measuring the Doppler frequency shift of a radio signal transmitted from ground stations to the satellite. The precision of the orbits is validated utilizing the altimeter crossovers. The altimeter crossover differences are on average less than 5 cm, which also constrains the across-track and along-track satellite positions.

In the PSI procedure, orbital errors that are spatially correlated are removed in the APS estimation, provided that they are uncorrelated in time. However, from Fig. 6.27 it is obvious that the phenomenon under investigation has a component that is correlated in time. The estimated PS velocities deviate $\sim$15 mm/year over 100 kilometers. Figure 6.28 shows the change in parallel baseline between near and far range caused by radial and across-track orbital errors. A spatial trend of $\sim$15 mm/year between near and far range implies a change in parallel baseline of $\sim$45 mm in three years. If the radial error stays within the 5 centimeter limit, this implies that the across-track error has to be $\sim$30 cm to cause a temporal difference in parallel baseline of $\sim$45 mm. This is an unlikely scenario, since the radial orbit errors of Envisat have been estimated at 3 cm RMS (Doornbos and Scharroo, 2004), which constrains the across-track and along-track position of Envisat. Moreover, a correlation between the moments of orbit manoeuvres and the development of the spatial trend in time in Fig. 6.27 could not be found.

Additionally, the influence of another gravity model for precise orbit determination has been investigated. The Envisat orbits have been computed using the orbits that are based on the EIGEN-GRACE01S and the EIGEN-CG03C gravity models

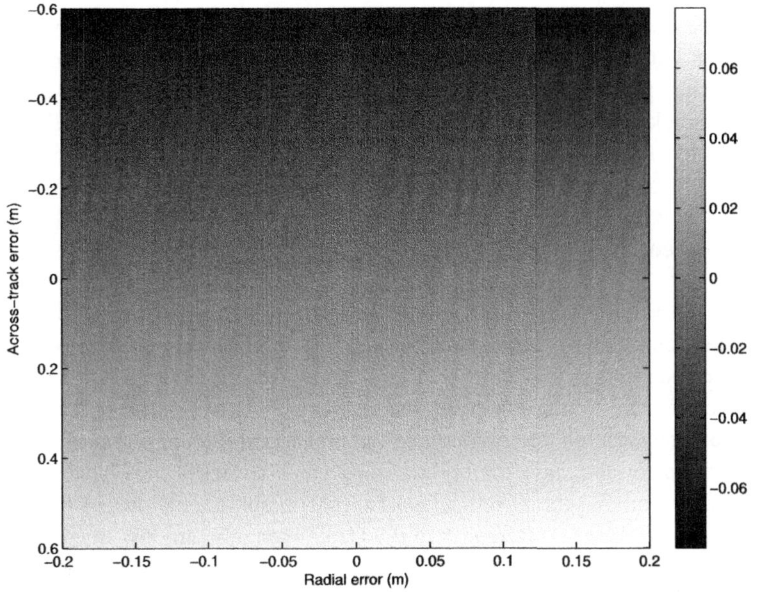

Fig. 6.28 Difference in parallel baseline between near range and far range caused by radial and across-track orbit errors (m). These orbit errors have to be interpreted as the radial and across-track components of the orbit error vector that is superposed on the baseline

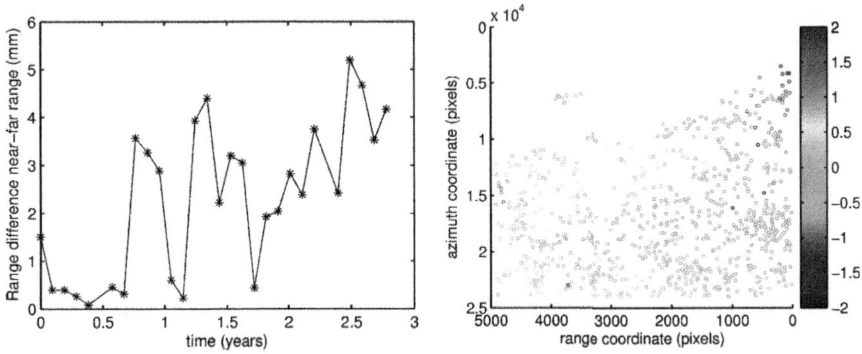

Fig. 6.29 The across-track range difference from near to far range (left) and the difference in PS velocity estimates (right), obtained from using the EIGEN-GRACE01S and EIGEN-CG03C orbits. This shows that the difference in PS velocities (1–2 mm/year) are of the same order as the range differences imposed by the across-track orbital differences (~1.2 mm/year)

in the period from December 2003 to November 2006. The variation in the radial component is below ~1–2 cm. The across-track component however, shows a systematic deviation in time of ~4 cm in 3 years, see Fig. 6.29. This corresponds with a range difference of ~4 mm in 3 years. Hence, the difference in gravity model explains the difference in PS velocity estimates of 1–2 mm/year in Fig. 6.29.

It can be concluded that the characteristics of across-track orbit errors when using different gravity models correspond with the time development of the spatial trends in the Envisat phase observations as depicted in Fig. 6.27. However, the magnitude of across-track errors that are required to explain the spatial trends in the Envisat observations are unlikely to occur, due to the radial precision (3 cm RMS).

6.3.2.3 System Parameters

Errors in the Envisat system parameters have been investigated as well to explain the spatial trends. This section addresses the influence of timing errors and an error in the range sampling rate (RSR).

The range time to the first pixel and the first pixel azimuth time can be tuned using the coregistration polynomial and the precise orbits. For a grid of radar coordinates in master and slave the optimal range and azimuth times have been determined by minimizing the misclosures of the geographic locations on the ellipsoid (the topographic height differences in the Groningen area are less than 30 meters). It turned out that two acquisitions have relatively large timing errors. These timing errors introduce a systematic effect in the reference phase. Figure 6.30 depicts the range difference between near and far range that is caused by the timing errors. The timing errors can introduce 1–2 fringes from near to far range. It can be seen from Fig. 6.31 that the outlier characteristics (see Fig. 6.24) of the timing errors affect the PS velocity estimates. However, they are not large enough to create a systematic trend in the PS velocities of ~15 mm/year from near to far range.

Furthermore, the effect of a constant error in the range sampling rate (RSR) has been investigated. The effect of a different RSR causes random changes in the range distance, see Fig. 6.31. Hence, the RSR error needs to be high in order to cause the observed trend in the PS velocities: ~0.2 MHz.

Summarizing, the cause of the spatial trends in the Envisat phase observations cannot be unambiguously identified. The spatial trends resemble errors that are

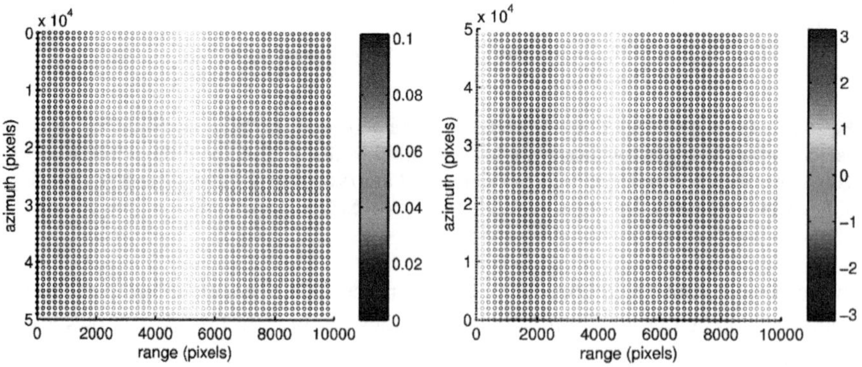

Fig. 6.30 Effect of timing errors over the extent of an interferogram: (left) errors in the parallel baseline (m) and (right) errors in the reference phase (wrapped, in radians)

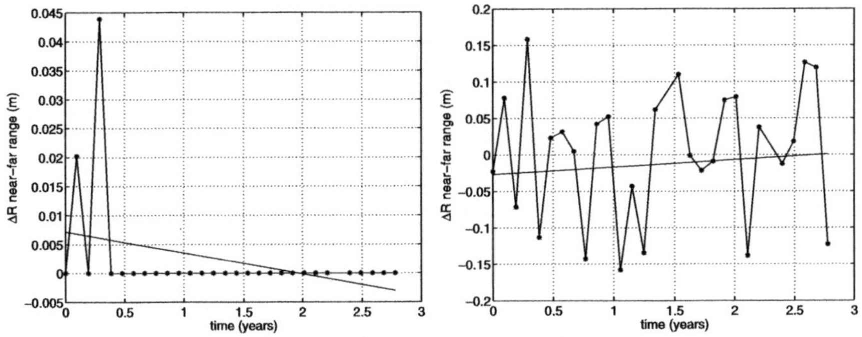

Fig. 6.31 Left: the effect of the Envisat timing errors on the range difference between near and far range. This implies that timing errors can cause systematic effects in the PS velocities, but that the magnitude is not large enough to cause the spatial trends that have been observed. Right: effect of an error of -0.2 MHz in the Range Sampling Rate (RSR)

caused by orbital inaccuracies. However, the magnitude of the spatial trends in the PS velocities ($\sim$15 mm/year from near to far range) cannot be explained by orbit errors within their precision bounds (Doornbos and Scharroo, 2004). Timing errors are too small and the RSR has to be adapted significantly to obtain the observed spatial trend in the PS velocity estimates. Moreover, changing the system parameters does not lead to the trend in time depicted by Fig. 6.27.

6.4 Multi-track Analysis

The precision of the Groningen PSI velocity and height estimates have been evaluated in Sect. 6.3. Since the PSI system of equations is not redundant and ambiguity resolution success rates of 1 can not be guaranteed (see Sect. 4.2), Chap. 5 has proposed multi-track datum connection for reliability assessment. Since six independent overlapping ERS tracks observe the Groningen subsidence area in the same period, redundancy is introduced. Besides a reliability assessment, the multi-track datum procedure performs the integration of PSI parameter estimates in a unified radar datum. This section shows the results of the multi-track datum connection of the ERS tracks that cover the Groningen area.

6.4.1 Datum Connection

For the datum connection of the six overlapping ERS tracks, the PS with the highest velocity estimate precision were selected. For this selection, the covered area was subdivided in grid cells of 500 meters. Subsequently, a spatial datasnooping procedure was applied to remove outliers. This procedure decreases the computation time of the datum connection parameters. This does not imply that PS are permanently removed from the dataset. The datum transformation estimated from this reduced set of PS can be applied to the original PSI results.

After conversion to the common radar datum defined by the master track (track 487), the datum connection procedure has been applied to the PSI estimates, see Sect. 5.2. For the velocity and displacement estimates, the alternative hypothesis of translation plus trend in range and azimuth directions has been evaluated. The reason for the need of including range and azimuth dependent transformation parameters stems from possible unmodeled residual atmospheric or orbital errors and undetected unwrapping errors that propagate over a large spatial extent.

Datum connection has resulted in a consistent set of PSI estimates in the reference system of the master track, covering the entire northern part of the Netherlands and a part of Germany, see Fig. 6.32. All subsiding areas due to gas extraction can be distinguished. Since standard adjustment and testing techniques have been applied, quality measures for the estimated transformation parameters can be deduced. Due to the large redundancy, the precision of the transformation parameters is high: the translations have a precision of 0.1–0.2 mm/year; the trend parameters in range and azimuth 0.1–0.3 mm/year over a 100 km distance, see Table 6.3. Approximately 70% of the velocities in a PS cluster have a standard deviation lower than 1 mm/year after datum connection (see Fig. 6.33), and are spatially consistent.

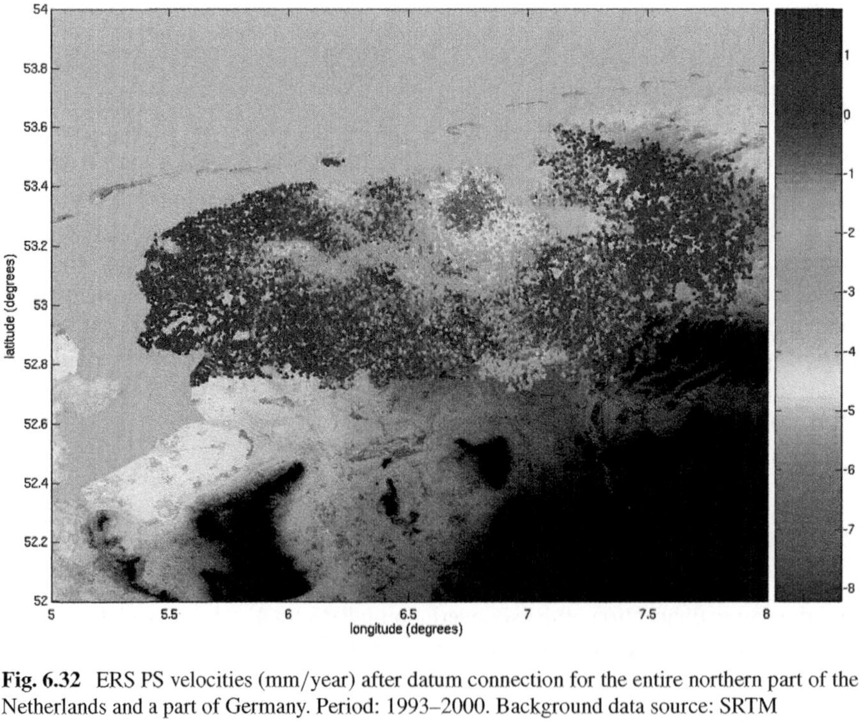

Fig. 6.32 ERS PS velocities (mm/year) after datum connection for the entire northern part of the Netherlands and a part of Germany. Period: 1993–2000. Background data source: SRTM

Table 6.3 Precision of PS velocity transformation parameters: a translation (t_0) plus azimuth and range dependent factors (t_ξ, t_η). The master track is track 487

Parameter	σ_{t_0} (mm/yr)	σ_{t_ξ} (mm/yr)/100 km	σ_{t_η} (mm/yr)/100 km
$v^{258,487}$	0.09	0.21	0.17
$v^{215,487}$	0.20	0.25	0.22
$v^{151,487}$	0.10	0.24	0.20
$v^{380,487}$	0.08	0.12	0.14
$v^{108,487}$	0.11	0.18	0.17

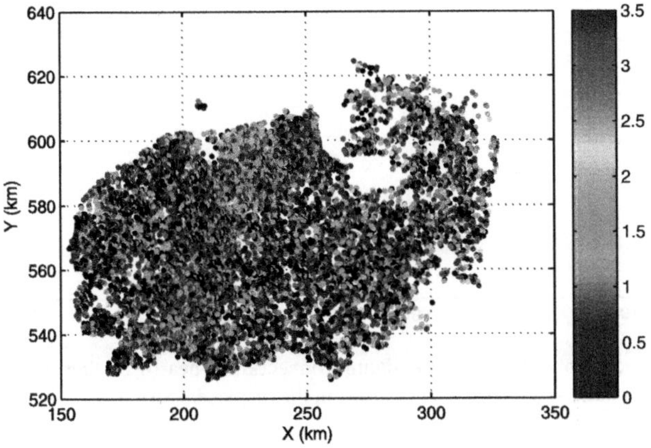

Fig. 6.33 Standard deviations of PS velocities per multi-track cluster after datum connection (mm/year). Approximately 70% of the velocities in a PS cluster have a standard deviation lower than 1 mm/year after datum connection

After datum connection, the PSI results are mutually consistent in the reference system of the master track. However, the reference system of the master track itself can still contain a small systematic component due to unmodeled residual effects. The standard deviation of the five trend estimates in range and azimuth direction are respectively 2 and 1 mm/year over a full SLC extent (100 km). Therefore, it can be concluded that a trend of several mm/year over a 100 km distance can be present in the PSI results after datum connection. Theoretically, this trend cannot be unambiguously addressed to either real deformation signal or unmodeled residual components in the PSI estimation. Since the latter is most likely, the deformation estimates obtained after data connection can be integrally corrected, as long as the error bounds are clearly defined.

Datum connection not only increases the spatial sampling of the deformation signal of interest, but also the temporal sampling is significantly higher. One PS is viewed by a maximum of 4 tracks, that are distributed over the repeat interval of 35 days, with intervals of 7–12 days.

In a similar way as demonstrated for the PS velocities, the PS displacements have been connected in a common radar datum, including corrections for residual

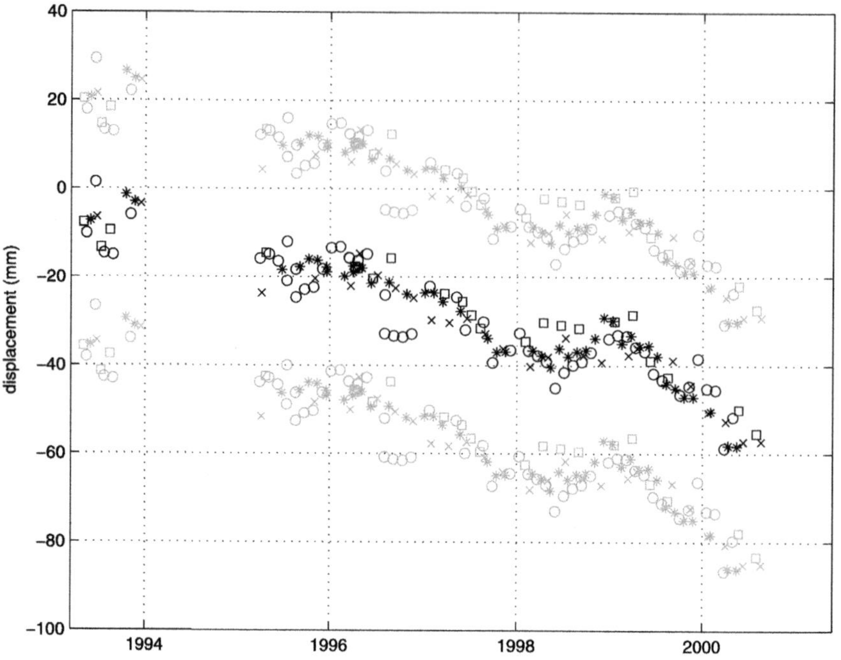

Fig. 6.34 Displacement time series after datum connection of nearby PS that are viewed from four tracks. The grey markers depict the ambiguities in the displacements

components in the reference track. A complicating factor are the different acquisition times per track. To avoid any assumptions on the temporal deformation model, displacement observations were linearly interpolated. This has resulted in a series of displacement fields with their corresponding transformation parameters through time.

Figure 6.34 shows the displacements of a cluster of neighboring PS (mutual distance less than 500 meters) that have been observed by four overlapping tracks. The integration of displacements from multiple tracks increases both the precision and the reliability. The increase in subsidence rate after 1996 is confirmed by all four tracks.

Besides the deformation estimates (velocities, displacements), the topographic heights have been integrated in the datum connection as well. Due to different reflection types (directly from a roof top, or a double-bounce that refers to ground level), and the variety of man-made features, the heights of neighboring targets do not necessarily correlate. Therefore, only PS within resolution cell distance have been selected. Again, outlier detection has been performed to remove targets within resolution cell distance that refer to a different height level. The precision of the height translation parameters is listed in Table 6.4.

Figure 6.35 shows the PS heights after datum connection and the topographic height that have been obtained from SRTM. Although PS heights are ellipsoidal heights and SRTM heights are orthometric heights, the relative PS heights and the

Table 6.4 Precision of the translation parameters of the PS height estimates in the datum connection procedure

Parameter	σ_{t_0} (m)
$\Delta h^{258,487}$	0.64
$\Delta h^{215,487}$	0.86
$\Delta h^{151,487}$	0.71
$\Delta h^{380,487}$	0.58
$\Delta h^{108,487}$	0.73

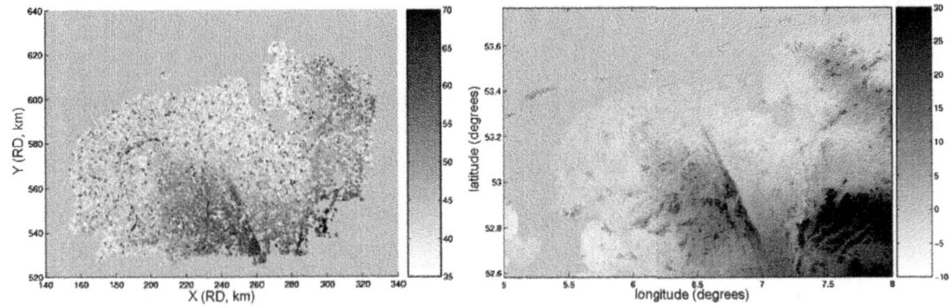

Fig. 6.35 PS heights (ellipsoidal, WGS84) and SRTM heights (orthometric, NAVD88)

relative SRTM heights agree visually. The differences are on average below 5 meters. They are caused by the accuracy of the SRTM heights, the accuracy of the PS heights, and the spatial differences between heights on the geoid and the ellipsoid. Furthermore, height differences can be caused by sidelobe observations and range sub-pixel inaccuracies in the PS results, see Sect. 4.2.

6.4.2 Displacement Vector Decomposition

The deformation estimates of the PS clusters after datum connection can be further decomposed into horizontal and vertical displacements. Each PS cluster contains at least two and maximum four PS that originate from different viewing geometries (ascending, descending, and adjacent). For a cluster of two PS, a decomposition into a vertical component and one horizontal component along a certain viewing direction is possible. For a cluster of more than two PS theoretically a decomposition into vertical, east and north components is possible. However, due to the acquisition geometry, the precision of the north component is lower than the precision of the east component (Wright et al., 2004).

In the Groningen study, a quadtree decomposition of the subsidence signal has been applied to increase redundancy. The PS deformation estimates within each quadtree grid cell have been decomposed into a vertical and a horizontal component along ascending look direction (Hanssen, 2001). Figure 6.36 shows that local horizontal movements for the main Groningen subsidence bowl are 2–3 mm/year

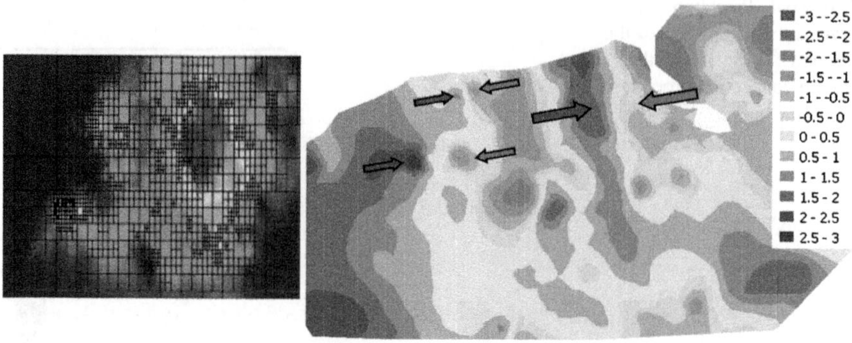

Fig. 6.36 Quadtree decomposition and interpolated horizontal PS velocities (mm/year) along ascending look direction

Table 6.5 Approximate Groningen reservoir parameters for the period 1993–2003

compaction coefficient (bar^{-1})	c_m	$0.72 \cdot 10^{-5}$
Poisson's ratio	ν	0.25
reservoir thickness (m)	H	150–220
pressure decrease in 10 years (bar)	ΔP	36

towards the center of the bowl. This local effect is also visible for some of the smaller subsidence bowls. Although the reliability of the horizontal components needs to be further investigated, both magnitude and direction approximately correspond with the theoretically predicted values (Geertsma, 1973b). The analytical model from Geertsma (1973b) has been evaluated on a simplified representation of the Groningen gas field as a disc shaped reservoir with a diameter of 30 kilometers at 3 kilometers depth. Average values were taken for Poisson's ratio, reservoir thickness and the compaction coefficient in the period 1993–2003, see Table 6.5.

Figure 6.37 shows that the maximum expected horizontal movements are around 3 centimeters in a ten year period, which corresponds with the magnitude of the horizontal PS velocities of 2–3 mm/year. The geophysical background of the spatial subsidence components and their estimation using PSI is further elaborated in Ketelaar et al. (2008a).

For Envisat, the PSI estimates of only one track are available. This track covers a time period that does not overlap with the ERS PSI estimates. Hence, a multi-track datum connection procedure cannot be performed. This does not imply that the Envisat results are unreliable for the estimation of subsidence due to gas extraction. Redundancy is introduced by the spatial sampling of the Groningen subsidence bowl, that has an extent of 30 kilometers. The spatial density of the Envisat PSI estimates will be demonstrated in the comparison with leveling displacements along profiles that cross the subsidence bowl, see Sect. 7.2.

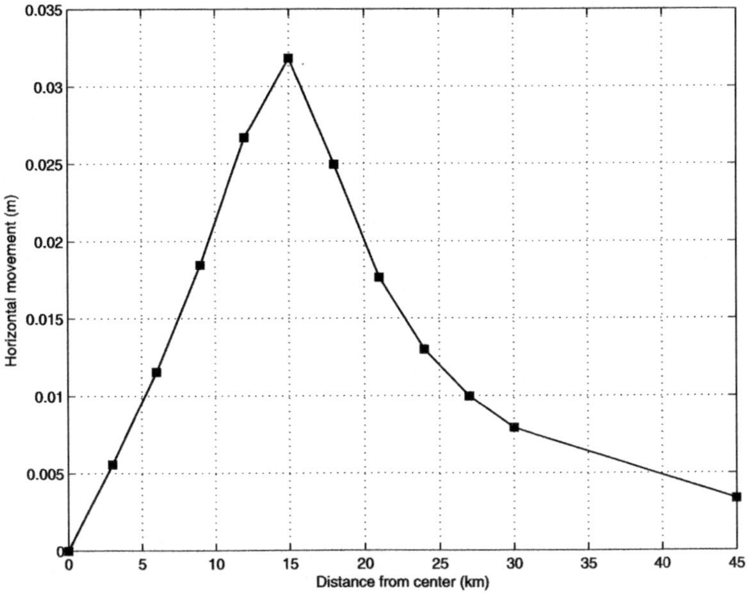

Fig. 6.37 Predicted horizontal movements in 1993–2003 for the Groningen subsidence bowl. This shows that horizontal movements are maximum ~3 mm/year

The multi-track datum connection of the six ERS tracks has shown that velocity and deformation estimates from multiple independent tracks can be integrated. The standard deviation of multi-track velocity estimates is less than ~1 mm/year. Both the precision and the reliability of PSI as a measurement technique have been quantified in Sect. 6.3 and Sect. 6.4. Now, Sect. 6.5 will focus on the estimation of subsidence due to hydrocarbon production from the PSI estimates. Physical PS properties and a-priori knowledge on the spatio-temporal behavior of the deformation signal will be utilized to investigate the idealization precision of PSI for monitoring subsidence due to hydrocarbon production.

6.5 Idealization Precision for Deformation Monitoring

Since subsidence due to hydrocarbon production can be contaminated by other deformation regimes (shallow compaction, structural instabilities), the PSI deformation estimates require further interpretation, see Sect. 4.5.1. This section starts with an inventarisation of deformation regimes in the Netherlands. Subsequently, the use of PS characterization and the spatio-temporal behavior of subsidence due to hydrocarbon production is investigated to assess the idealization precision for deformation monitoring.

6.5.1 Identification of Deformation Regimes

The combination of soft soils in the shallow subsurface and the low subsidence rates due to gas and oil extraction requires the investigation of deformation regimes in the Netherlands. Brand (2002) addresses several deformation processes in the Netherlands in order to explain leveling benchmark movements. Leveling benchmarks are subdivided into well-founded underground benchmarks and benchmarks that are mounted in existing buildings with varying foundation types. Underground benchmarks are founded on Pleistocene sand layers (Fig. 6.38), which are considered stable. In the Netherlands there are approximately 500 underground benchmarks, and 30000 benchmarks mounted in buildings. Due to geological activity, also the underground benchmarks exhibit small movements of maximum ~0.1 mm/year (ibid.). Figure 6.39 depicts the geological structure of the shallow subsurface of the north-

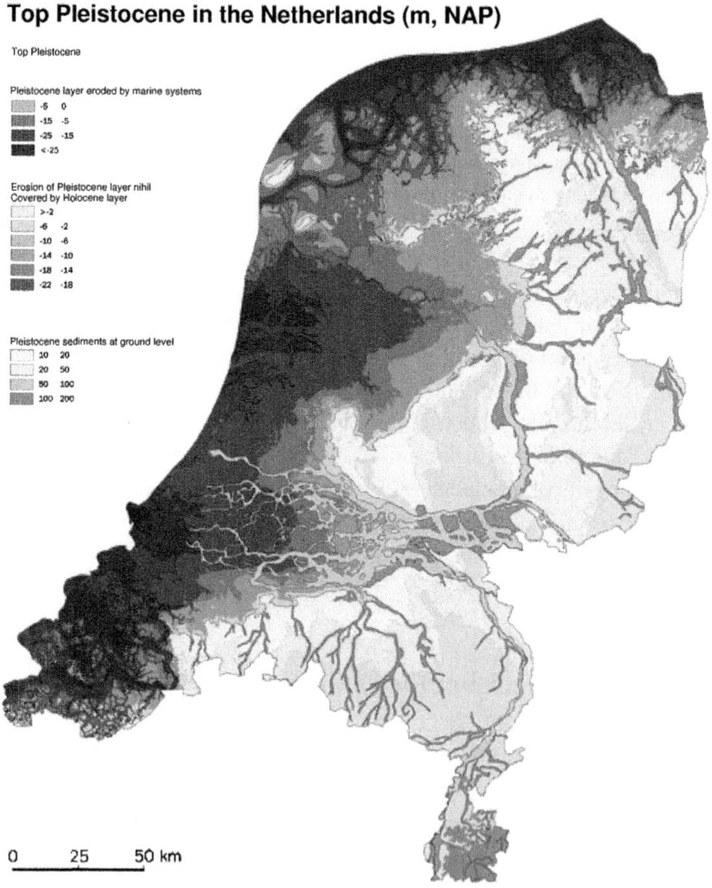

Fig. 6.38 The Pleistocene layer in the Netherlands (DINO, 2008). The top of the Pleistocene layer in the Dutch vertical datum (NAP, m) is depicted. Source: TNO Bouwen Ondergrond afd. DINO

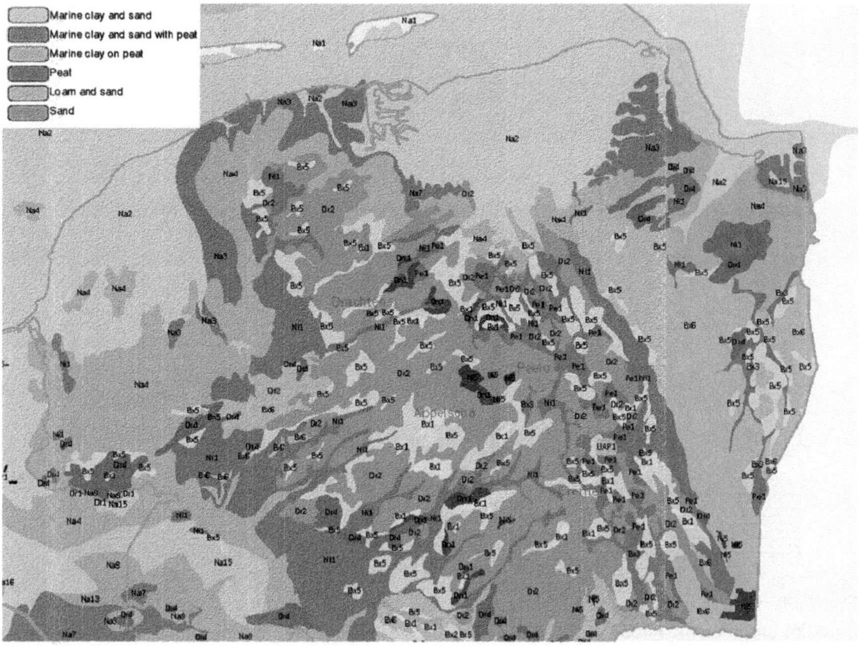

Fig. 6.39 Geological map of the northern part of the Netherlands (DINO, 2008; de Mulder et al., 2003). The subsurface consists of sand, peat and clay layers. Source: TNO Bouwen Ondergrond afd. DINO

ern part of the Netherlands. The shallow subsurface consists of sand, clay and peat layers.

6.5.1.1 Settlement of Foundations

The settlement of foundations is dependent on the type of foundation and the subsurface layers. The settlement can be caused by:

- the weight of the structure or building,
- negative pile friction (additional downward force due to settling subsurface layers),
- peat oxidation below the foundation,
- rotten piles.

The settlement of a foundation on sand is rather small and occurs in a short time period. A foundation on clay or peat causes settlement of a higher magnitude over a longer time period. The settling velocity decreases logarithmic in time. Special attention has to be paid to areas where peat oxidation might occur.

Figure 6.40 depicts the building foundation types in the Netherlands. Not all buildings have been founded on the stable Pleistocene layer. A common practice in Friesland and Groningen has been a foundation ‘op staal’: a shallow foundation

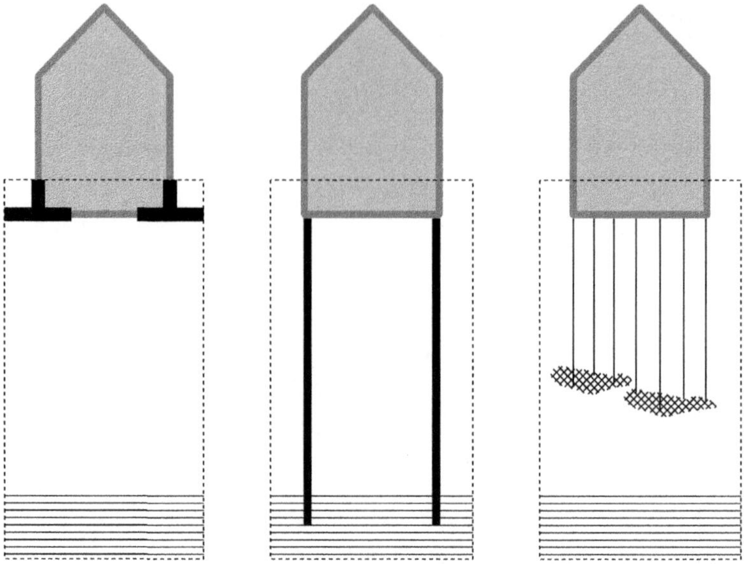

Fig. 6.40 Foundations in the Netherlands (CBG, 1987): (left) 'op staal' (shallow foundation), (middle) on poles mounted on the stable Pleistocene layer, (right) 'op kleef' (poles connected to stable parts in the shallow subsurface)

on the Holocene clay layers. Another type of foundation is 'op kleef': a collection of poles in the (shallow) subsurface layers, that connect the building to stable parts of these layers. The settlement of buildings has been monitored periodically by leveling the benchmarks that are mounted on these buildings. In Groningen, the majority of the benchmarks exhibit relative displacement rates within 1 mm/year (NAM, 1991; Hoefnagels, 1995). Figure 6.41 depicts a histogram of the benchmark displacement rates in the Groningen area before the start of gas extraction (NAM, 1991). Although the majority of the benchmark displacement rates are less than 1 mm/year, it has to be noted that areas can be indicated were all benchmarks exhibit slightly larger displacement rates, such as the area around Delfzijl (mideastern part of the Groningen subsidence bowl). Cheung et al. (2000) has detected maximum rates of −2.5 mm/year in Friesland, in areas with clay layers that are superposed on peat layers.

6.5.1.2 Shallow Subsurface Movements

Shallow compaction can occur as a result of the weight that is imposed on the layers, due to a decreasing groundwater level, or due to peat oxidation. The degree of compaction depends on the soil type: clay and peat layers are more affected than sand layers. Furthermore it is dependent on the depth of the layer: deeper layers have already been compacted and an additional pressure increase will have a reduced effect.

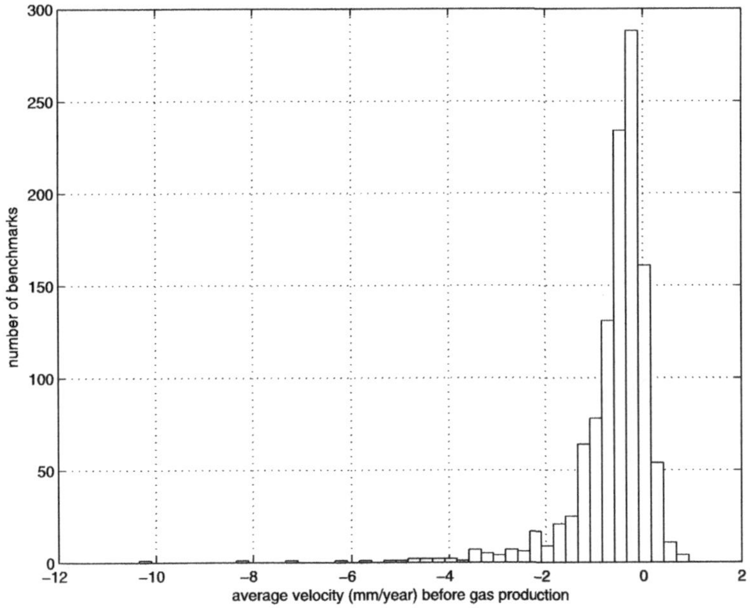

Fig. 6.41 Estimated benchmark displacement rates in the Groningen area before the start of gas extraction. The majority of the relative displacement rates are within 1 mm/year

The effect of groundwater level changes is demonstrated in the Delft corner reflector experiment, see Sect. 4.4. Five corner reflectors have been periodically leveled since March 2003. The corner reflectors are mounted in the shallow subsurface, ~40–50 cm deep. The corner reflector heights, that have a precision of 0.5–1 mm, vary seasonally with an amplitude of 1–2 centimeters. Groundwater tubes have been established in August 2005 near the four corner reflectors that were left. Measuring the groundwater heights has been incorporated in the leveling campaigns at the time of each satellite pass. Figure 6.43 shows that the seasonal behavior of the groundwater levels and the corner reflector heights is the same. Furthermore, it can be deduced that the amplitude of the groundwater level variations is a factor 10 higher than the amplitude of the corner reflector heights. If the groundwater level increases, the soil expands and the corner reflectors move upwards. If the groundwater level decreases in summer, the pore water pressure drops. This results in an increased effective stress, leading to compaction of the soil.

Since the magnitude of shallow compaction is potentially larger than the magnitude of subsidence due to gas extraction (cm/year versus mm/year), well-founded benchmarks are required to estimate subsidence due to gas extraction. Regarding the PSI technique, this implies that only observations with respect to well founded structures and buildings should be used.

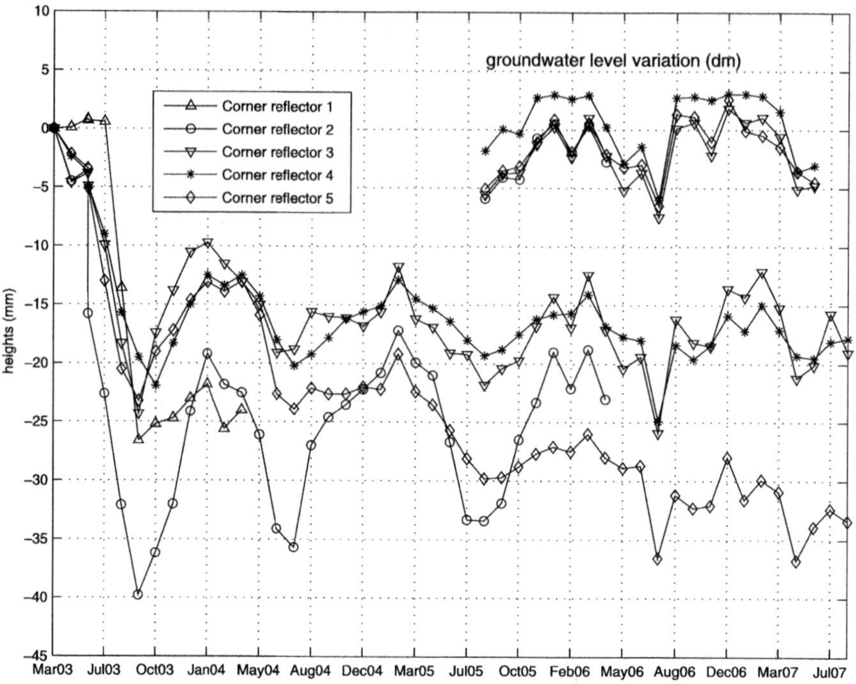

Fig. 6.42 Corner reflector heights (mm) since March 2003 and groundwater level heights (dm) since August 2005. The corner reflector heights show a seasonal amplitude of 1–2 cm. The corner reflector heights follow the groundwater level variations

6.5.1.3 Isostacy and Tectonics

Isostacy is the equilibrium of the lithosphere on the asthenosphere (the soft part of the earth's upper mantle). Movements to maintain or recover the equilibrium are very small and affect the entire lithosphere. They may for example be caused by the melting of a large ice cap. Since the last ice age, Scandinavia is still moving upwards and the Netherlands downwards.

Tectonics is concerned with the internal deformation of the lithosphere as a result of dynamics within the earth. Plate movements along faults may cause earth surface movements. Earth surface movements due to both isostacy and tectonics are expected to be maximum 0.1 mm/year (Brand, 2002).

6.5.1.4 Mineral Extraction

Besides gas and oil extraction, there are several salt mining areas in the Netherlands. Furthermore, until the 1970s, coal mines were operational in the southern province Limburg.

Subsidence at ground level due to hydrocarbon production has already been explained in Sect. 2.1.4. Section 6.5.2 will now specifically address shallow subsurface

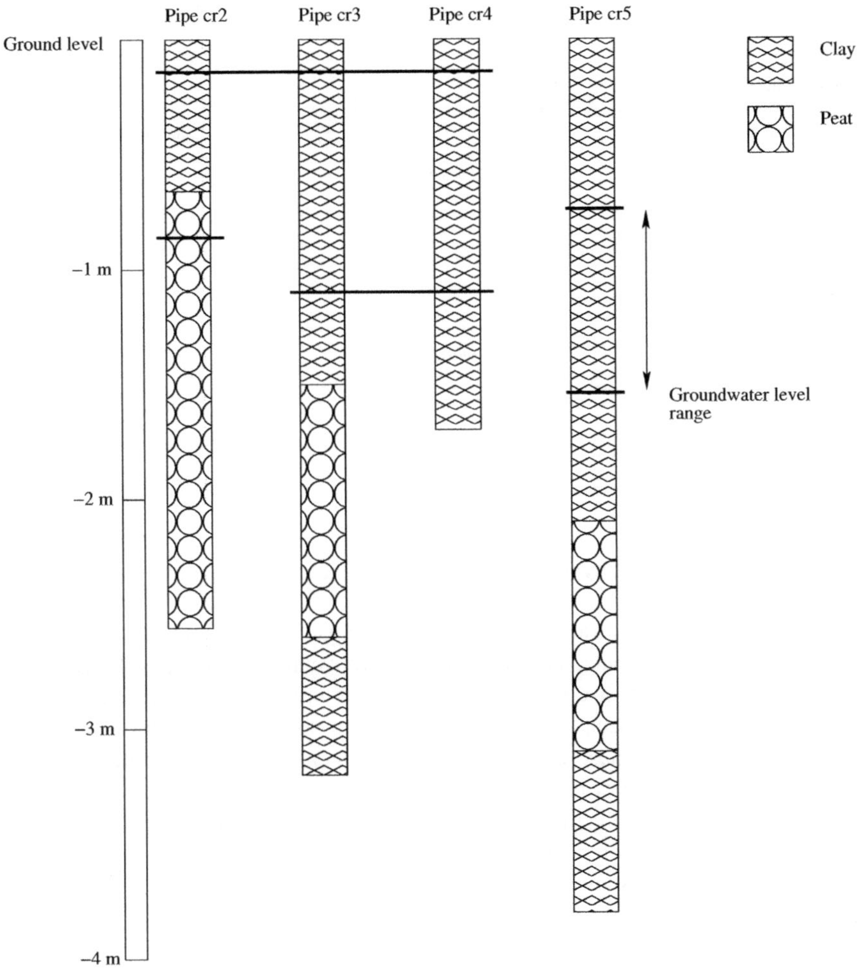

Fig. 6.43 Soil types (clay, peat) and ground water level variations per groundwater tube at corner reflector 2 (cr2), 3 (cr3), 4 (cr4), and 5 (cr5)

displacements in the Groningen area, that are superposed on the deep subsurface displacements.

6.5.2 *Shallow and Deep Subsurface Movements in Groningen*

Knowledge on shallow and deep subsurface movements in the Groningen area has been obtained in two ways: by shallow and deep observations wells, and by benchmark stability analyses. Both methods are successively described in this section.

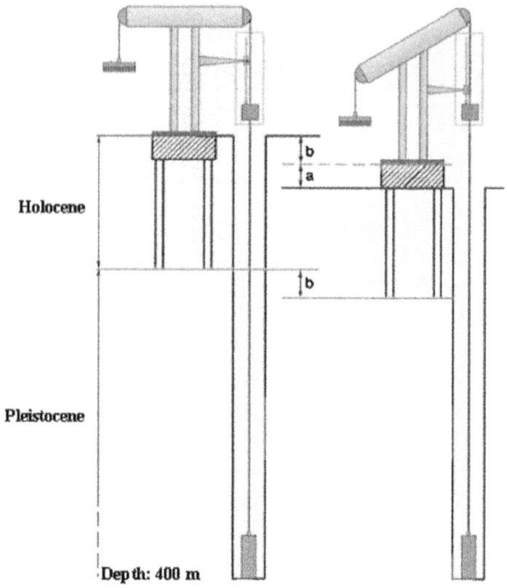

Fig. 6.44 Shallow observation well. The cable movement represents the shallow subsurface movements (de Loos, 1973; NAM, 2005)

6.5.2.1 Shallow and Deep Observation Wells

In the Groningen gas extraction area, in situ compaction measurements have been carried out to discriminate shallow subsurface movements from subsidence due to gas extraction. Reservoir compaction can be measured at 7 wells (originally 11), while 14 wells have been established to measure shallow compaction (de Loos, 1973; NAM, 2005). These compaction measurements verify the predicted subsidence and the behavior of the overburden.

Deep observation wells measure compaction due to gas extraction. The measurement targets for deep compaction are radio-active bullets that have been shot in the formation at regular distance (NAM, 2005). Their relative displacement is measured periodically by means of a gamma-ray detector.

Shallow compaction is measured by a cable installation in a shallow compaction well, see Fig. 6.44. This cable installation measures the movement of a cable that is held under constant tension by means of an anchor weight at the bottom of the well and a counterweight at the surface. The cable movement, that can be measured with sub-mm precision, represents the deformation in the layers between the bottom of the well and the surface. Natural compaction rates based on benchmark histories in Groningen are in the range of 0.5–8 mm/year (de Loos, 1973). The largest compaction occurs in the upper 50 m Holocene peat and clay layers. In the shallow compaction time series, the tidal influence is visible as well (ibid.). The highest magnitude of the oscillation due to tides is ~0.25 mm.

6.5.2.2 Benchmark Stability Analysis

To estimate subsidence due to gas extraction utilizing second order benchmarks (benchmarks with a shallow foundation), knowledge about possible benchmark movements due to other causes (natural subsurface movements, foundation instability and settlement of the buildings) is required.

Schoustra (2006) performs a benchmark stability analysis in the Groningen area that is based on geostatistics and physical properties. Benchmark stability analysis based on geostatistics utilizes the spatial correlation of benchmark movements. It assumes that benchmark movements due to shallow compaction exhibit a lower or no degree of spatial correlation. Both the SuRe methodology (see Sect. 2.3.4) and Kriging cross-validation (see Sect. 4.5.3.1) using benchmark heights in the Dutch vertical reference system (NAP) have been applied. In this way, 878 stable benchmarks (out of 2080) were selected that cover the entire Groningen subsidence area.

6.5.3 PS Characterization

To improve the idealization precision of PSI for the estimation of subsidence due to gas extraction, two options are available:

1. PS selection based on physical properties and reflection type (see Sect. 4.5.2),
2. the use of a-priori knowledge on the spatio-temporal behavior of subsidence due to gas extraction (see Sect. 4.5.3).

This section investigates the methodologies for PS characterization in the Groningen study: PS heights, reflectivity as a function of viewing geometry, and polarimetric observations.

Direct reflections from well founded buildings are assumed to be the most suitable targets to provide observations for the estimation of deep subsurface displacements. In this section, we try to isolate these PS from PS that originate from double-bounce reflections that refer to ground level. Subsequently, their deformation estimates are analyzed. It is expected that the PS that represent solely the deep mass displacements will exhibit the lowest displacement rates, since all deformation regimes superposed on the deep mass displacements contribute additional subsidence components. A selection of PS that represent only deep subsurface displacements should therefore result in a shift of the velocity estimates towards displacement rates of a lower magnitude. In this section, we try to quantify this shift to be able to make a statement on the usage of different PS types for the estimation of subsidence due to hydrocarbon production.

6.5.3.1 PS Heights

PS that represent direct and indirect reflections from buildings can be distinguished based on their height with respect to ground level (Perissin, 2006). In the northern part of the Netherlands, two case study areas have been selected: Groningen city and a rural area with scattered farms in the onshore part near the Waddenzee, see Fig. A.1. Sidelobe observations have been removed, since their height estimates

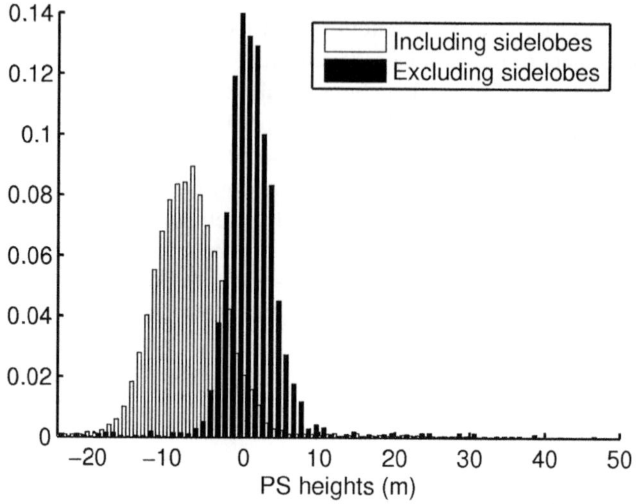

Fig. 6.45 PS heights before sidelobe removal (white) and after sidelobe removal, corrected for the reference PS height (black). Since sidelobes have incorrect height estimates, the distribution of the PS height after sidelobe removal is smaller

are incorrect, see Sect. 4.2.2. Figure 6.45 shows the PS height histogram before and after sidelobe removal. It shows that the PS heights have a much smaller distribution after rejection of these false PS targets. Furthermore, it can be seen that the peak of the height histogram before sidelobe removal is located at −10 m. Assuming that the height histogram peak corresponds with ground level (ibid.), this implies that the reference PS is located at +10 m above ground level. The height histogram after sidelobe removal has been shifted; the ground level now corresponds with a height of 0 m.

If the area of interest would be flat, PS heights with respect to ground level can be determined from the height histogram. However, even in the northern part of the Netherlands, ground level height variations of several meters cannot be neglected if ground level reflections have to be discriminated from reflections that refer to the top of a building. Hence, the following options for the determination of heights above ground level have been considered:

1. local PS height histograms,
2. SRTM heights,
3. laser altimetry heights.

The first method works if small areas can be considered flat (height differences less than one meter). It creates height histograms as depicted in Fig. 6.45 for small areas. In each of these areas, the peak of the histogram is assumed to indicate the ground level. Subsequently, the PS heights above ground level can be determined.

The second option considers the SRTM heights, that are sampled each 3 arc seconds. Rodriguez et al. (2005) perform an accuracy assessment of the SRTM data. For Europe, the height errors are in the order of 6–8 m. This is not accurate enough

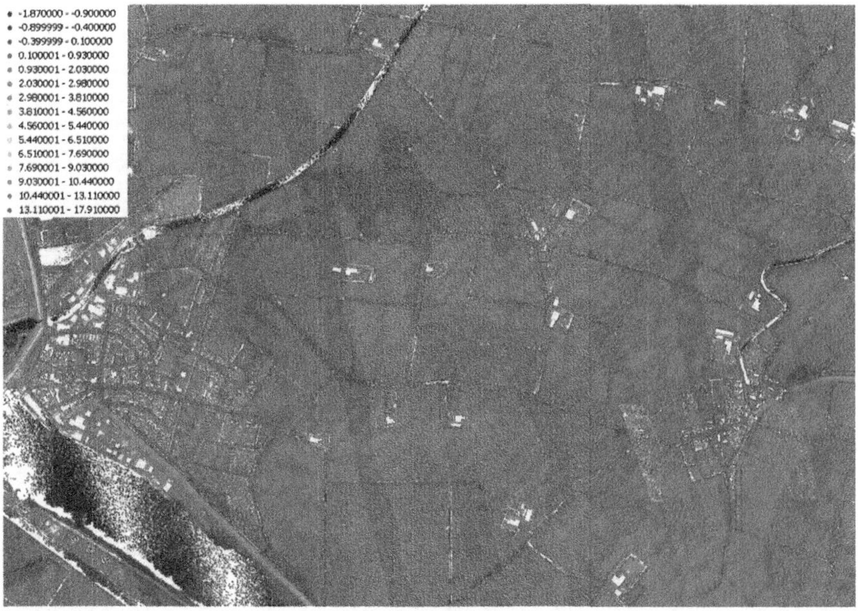

Fig. 6.46 AHN heights in a rural area (m), © AHN-www.ahn.nl. This data product aims to represent the ground level heights and hence the elevated targets, i.e. buildings, have been removed (white spots)

to distinguish reflections from the average building height in the Netherlands from ground level. Moreover, since the SRTM heights stem from radar interferometry, they do not necessarily represent ground level.

The last option utilizes the Actual Height model of the Netherlands (AHN, 2008). In the period 1996–2003, airborne laser altimetry data has been acquired covering the Netherlands. The precision of the heights is 10 centimeters, with a density of 1 point per 16 m^2 (ibid.). AHN products can be subdivided in filtered point heights at original locations, and interpolated point heights at various postings (5, 25, 100 meters). The filtered point heights refer to ground level, except for urbanized areas larger than 1 km^2 (ibid.). This implies that in the areas where most PS are found, the AHN point heights are not at ground level. The scattered white spots in Fig. 6.46 show that the buildings have been successfully filtered out in the rural area, whereas in Groningen city, the AHN heights form a Digital Elevation Model of the city, see Fig. 6.47. The AHN height histograms of the rural area and Groningen city are depicted in Fig. 6.48.

The first method, that uses the local height histograms, has been chosen as the preferred method to estimate heights above ground level. This method works for both rural and urban surroundings, where local areas can be considered flat. After computation of the PS heights above ground level, all PS with a height larger than 5 meters have been selected. Subsequently, the velocity estimates of the selected PS have been evaluated. Since the selected PS are likely to represent direct reflec-

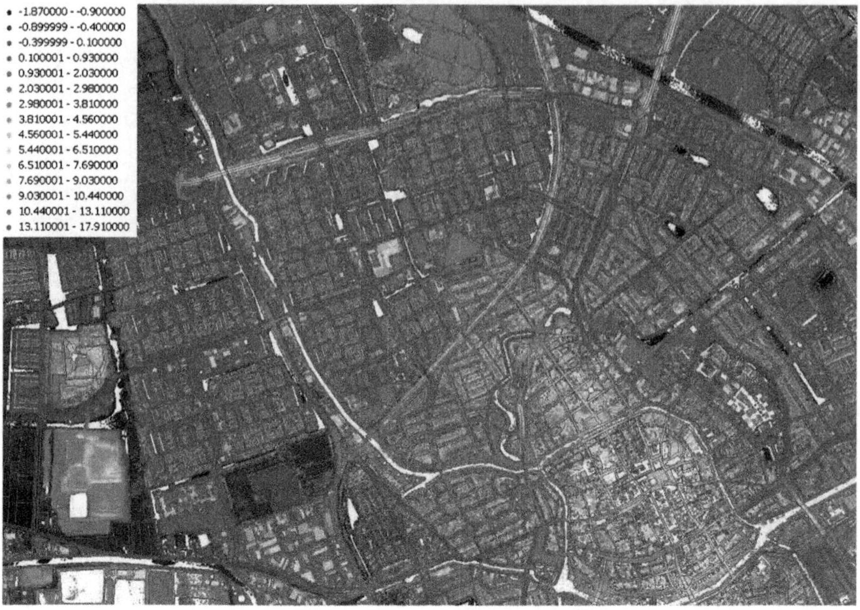

Fig. 6.47 AHN heights in Groningen city (m), © AHN-www.ahn.nl

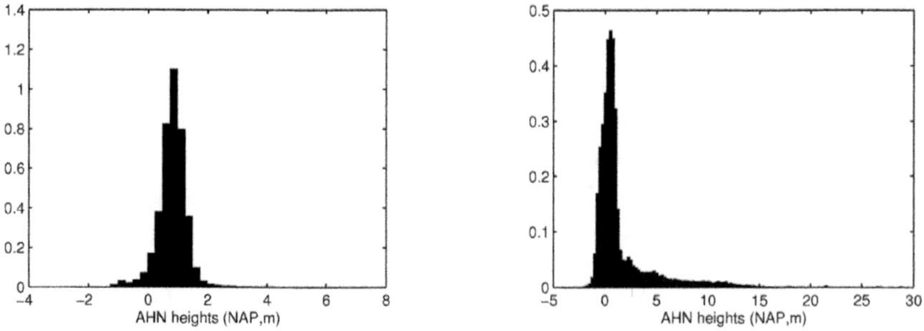

Fig. 6.48 AHN height histogram in the case study areas: a rural area (left) and Groningen city (right)

tions from (stable) buildings, the magnitude of the displacement rates is expected to be lower. This is shown in Fig. 6.49, which depicts the PS velocity histograms before and after selection on PS height above ground level in both case study areas. All PS velocities are relative, and stem from PS processing; a selection based on correlation with neighboring PS has not been performed. As expected, the velocity histogram shifts to the right after selection. Furthermore, the histogram's shape becomes smaller, meaning that the standard deviation of the velocities decreases (hypothetically due to less additional autonomous components).

Although a change of histogram shape is visible after selection on PS height, the shift in PS velocity is not more than 0.5 mm/year and the location of the histogram

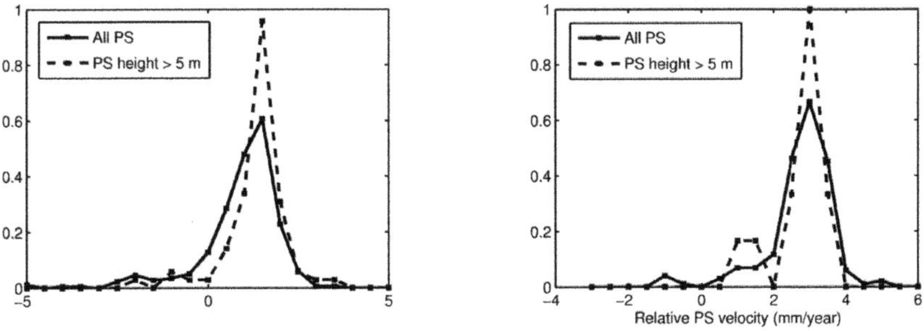

Fig. 6.49 PS velocity histograms for all PS and a selection of PS that are more than 5 meters above ground level: (left) Groningen city and (right) Waddenzee onshore

peak stays the same. A hypothesis to explain this minimal changes after selection on PS height is that PS reflections are related to well-founded structures, irrespective of the type of the reflection. Another explanation may be that subsidence due to shallow compaction and structural instability has a less smooth and less linear behavior than ongoing subsidence due to deep compaction. As a result, PS representing deep compaction may have a higher coherence. Furthermore, PSI is a relative technique, and hence cannot detect shallow displacements if they are equal for the entire area.

The same analysis of the velocity histograms has been performed before and after PS selection based on reflectivity as a function of viewing geometry (Ketelaar et al., 2006), see Sect. 4.5.2.2. For the Groningen descending main track, in total 106 acquisitions were available with a Doppler range of 9050 Hz, which corresponds with a squint angle variation of 5.3 degrees. The range in perpendicular baseline is 2215 m, causing a viewing angle range of 0.13 degrees. The same conclusion was drawn as for selection based on PS height above ground level: the PS velocity distribution is more concentrated in the stable domain, but the shift before and after PS selection is not significant (<0.5 mm/year).

It has to be stressed that the two case studies do not necessarily represent the entire area that is affected by subsidence due to gas extraction from the Groningen field. It is recommended to verify the effect of PS selection in areas that are classified according to the shallow geology, as depicted in Fig. 6.39.

6.5.3.2 Alternating Polarization

Besides PS heights above ground level and PS reflectivity as a function of viewing geometry, dual polarization observations can be utilized for PS characterization. In Sect. 4.5.2.3, it has been shown that pixels with a high amplitude in an HH-VV AP image have two distinct responses that have a phase difference of π.

Unfortunately, due to conflicts with commercial users, only one AP image has been acquired over the Groningen subsidence area, from the ascending track 487. The available AP image from track 487 has been acquired on the 1st of January 2006 and has a baseline of 1070 meters with respect to the ERS master. This means

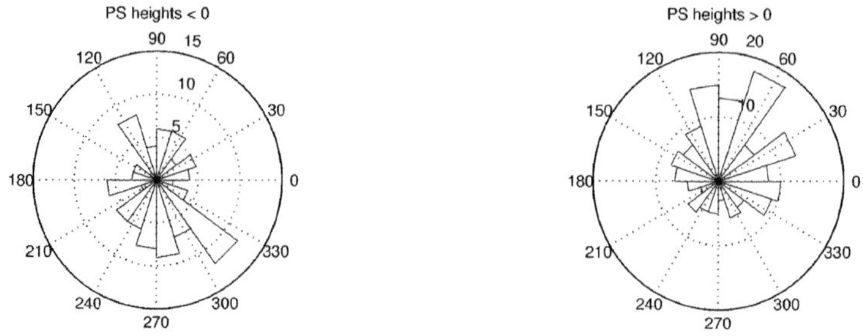

Fig. 6.50 Rural area: HH-VV phase differences for PS heights below and above ground level

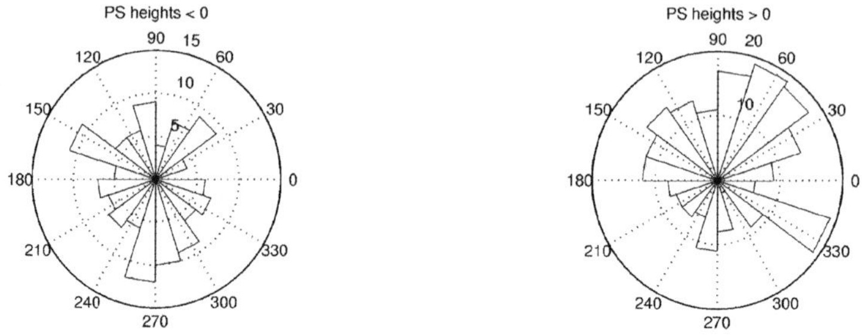

Fig. 6.51 Groningen city: HH-VV phase differences for PS heights below and above ground level

that the ERS and Envisat spectra do not overlap, and the AP information can only be successfully linked to Persistent Scatterers that act like ideal point targets.

Coregistration of the Envisat AP images with the ERS master scene has been performed similarly to the multi-track approach. Based on the orbits, an initial coregistration polynomial has been estimated, which has subsequently been refined using the point fields. For the Envisat AP image, this point field is reconstructed by pixels that exceed a certain amplitude threshold. Because of the decreased azimuth resolution of the AP image, it is complicated to use the amplitude information.

By means of the AP data, it is investigated if PS with a higher height above ground level are odd-bounce scatterers (most likely specular reflections). The same case study areas have been used (rural area and Groningen city). The PS heights above ground level have been determined using local height histograms.

Figures 6.50 and 6.51 show the HH-VV phase differences for targets above and below ground level. PS targets above ground level have a polarimetric phase difference that is concentrated around $\pi/2$; the uncertainty is considerable. To optimize the probability that PS targets are indeed located above ground level, the uncertainty of ground level determination has to be taken into account. Considering ground level targets and elevated targets as different height regimes, the height distribution is a multi-modal distribution. Figure 6.53 shows the histogram fit of the ground level deduced from the PS height estimates in Groningen city. The standard deviation of the best fit is approximately 2.5 meters. Translating that into a 95% confidence interval, the critical height value is 5 meters. Figure 6.52 shows that most targets that are clas-

Fig. 6.52 PS targets before (left) and after (right) selection on height above ground level (5 meters). The selected PS targets correspond with buildings; the PS at lower heights are deselected. Map: © 2007 Google Earth, Image © 2007 Aerodata International Surveys, © 2007 Europa Technologies

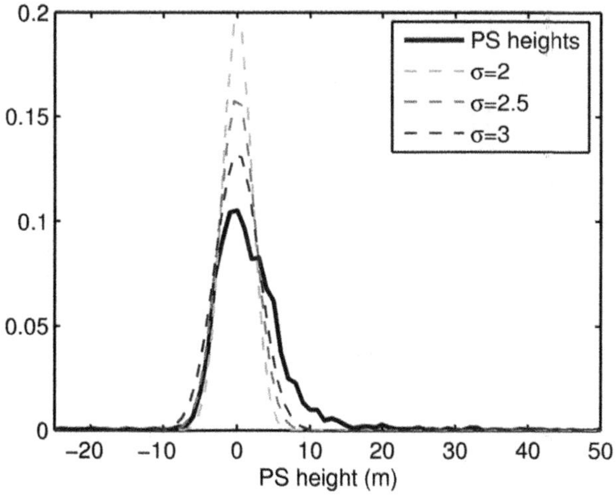

Fig. 6.53 Groningen city: ground level histogram fit using the PS height estimates. The best fit corresponds with a standard deviation of ground level heights of 2.5 m

sified as 5 meters above ground level are connected to buildings. From Fig. 6.54 can be deduced that the majority of these PS can be classified as odd-bounce reflections, most likely specular reflections.

It can be concluded that AP phase observations can aid the distinction of odd-bounce reflections from elevated targets (buildings) in the Groningen area. However, when considering the velocity histograms before and after PS selection based on PS height, it can be concluded that the effect of PS characterization is not significant in the case study areas. Here, the majority of the PS seem to refer to well-founded structures that subside due to the common deformation regime of gas extraction. However, the PS characterization tools should certainly be considered for interpretation in areas with considerable (variations in) shallow compaction.

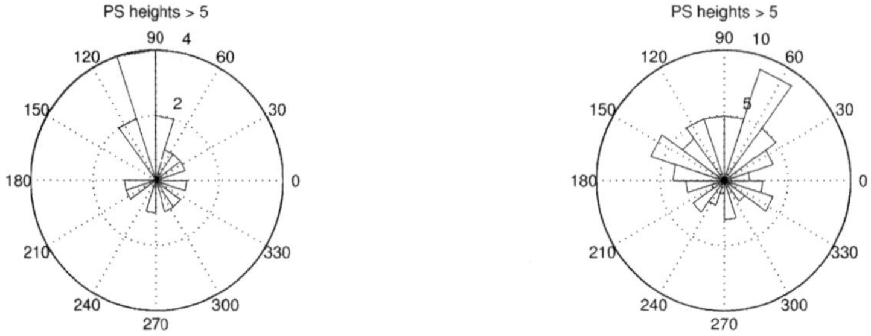

Fig. 6.54 HH-VV phase differences for PS height estimates that are more than 5 meters above ground level for (left) the rural area and (right) Groningen city

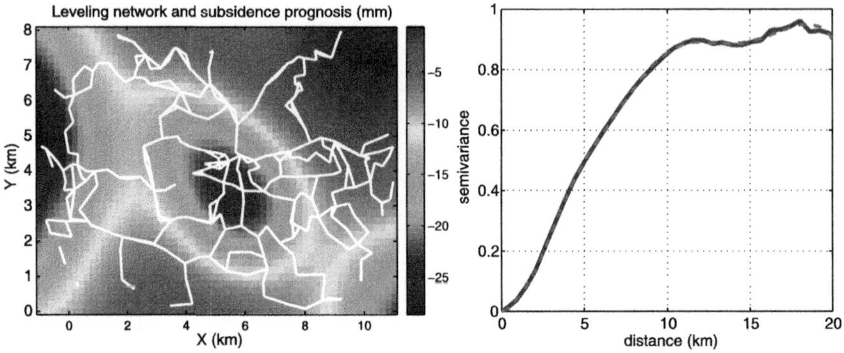

Fig. 6.55 Left: subsidence prognosis of the Rotterdam gas field (mm), including the leveling trajectories (white). Right: variogram of the subsidence prognosis

6.5.4 On the Use of A-priori Knowledge on the Deformation Signal

Besides PS characterization, the use of a-priori knowledge on the spatio-temporal behavior can increase the idealization precision of PSI for deformation monitoring. This section shows an application of Kriging cross-validation and an application of the decomposition of PS displacements according to deformation regime (see Sect. 4.5.3.2). Both methodologies are demonstrated for the estimation of subsidence due to gas extraction from the Rotterdam gas field. The Rotterdam gas field is situated in the western part of the Netherlands. The subsiding area is small, $\sim$25 km^2, and displacement rates are only several mm/year. The discrimination between different deformation regimes is essential due to the low subsidence rates, combined with natural compaction due to the soft soils in the shallow subsurface.

6.5.4.1 Kriging Cross-Validation: The Rotterdam Casestudy

The Kriging weights for Kriging cross-validation have been determined based on the variogram of the subsidence prognosis, see Fig. 6.55. Subsequently, cross-validation

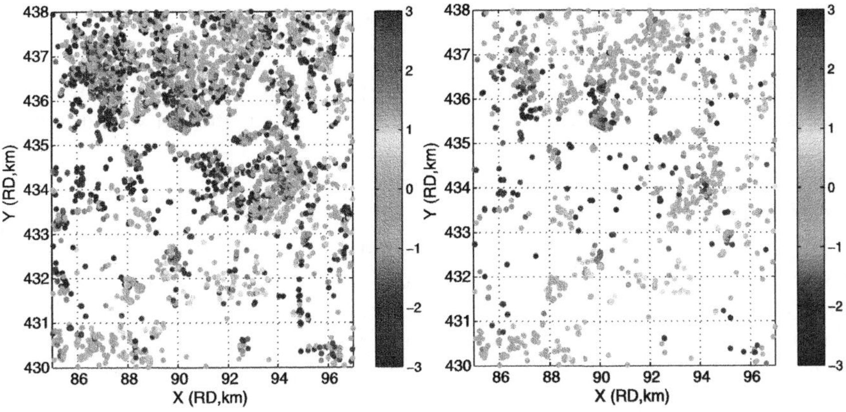

Fig. 6.56 PS velocities before (left) and after (right) Kriging cross-validation in the Rotterdam subsidence area. The depicted displacement rates have to be interpreted spatially relative

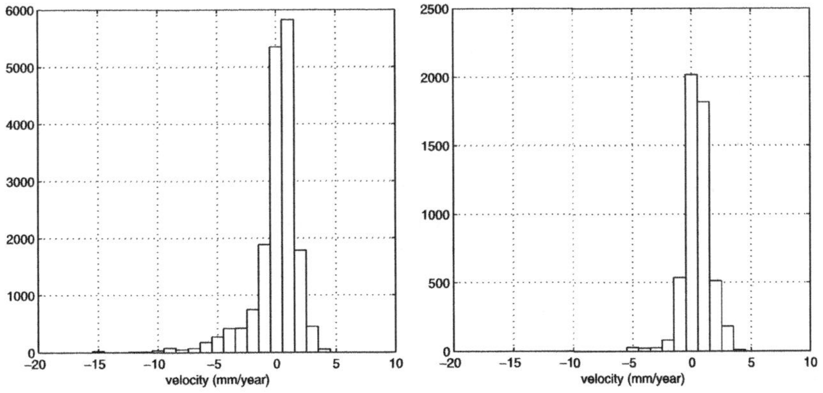

Fig. 6.57 Differences between PS velocities and prognosed velocities (mm/year), before (**a**) and after (**b**) Kriging cross-validation

has been applied to select the PS that exhibit similar displacement rates as their neighboring PS. Figure 6.56 shows the PS velocities before and after PS selection based on Kriging cross-validation.

Figure 6.57 depicts the differences between PS velocities and prognosed subsidence velocities before and after selection. It shows that the outliers have been successfully eliminated after cross-validation. Furthermore, it can be seen that the skew-symmetric left-hand tail of the original histogram has almost completely disappeared after PS selection. A hypothesis is, that the rejected PS contain additional autonomous displacement components caused by shallow subsurface movements or structural instabilities.

6.5.4.2 Displacement Decomposition: The Rotterdam Casestudy

This section demonstrates the estimation of displacement components in the Rotterdam area to improve the idealization precision of the estimation of subsidence due to gas extraction. For the estimation of subsidence due to gas extraction from the Rotterdam field, the separation of deformation regimes is essential due to the soft soils in the shallow subsurface.

PS selection using Kriging cross-validation removes PS that contain displacements due to gas extraction, but are contaminated by autonomous movements due to e.g., shallow compaction. Here, variance component estimation is applied to estimate the displacement components due the gas extraction from *all* available PS over the Rotterdam gas field (Ketelaar et al., 2004b). The methodology has been explained in Sect. 4.5.3.2.

Both leveling and PS displacements can potentially be subdivided into spatially and temporally correlated components and components that are temporally correlated only. The first category refers to the subsidence prognosis and its imperfections: it is referred to as the *model components*. The second category refers to individual point characteristics, such as foundation pressure and pile friction. This category consists of the *autonomous components*. For each deformation regime that is responsible for a part of the total PS displacement, the behavior is characterized by the stochastic model parameters: variance factor (magnitude) and correlation length.

The area under investigation has an extent of approximately 100 km^2, and has successively been evaluated with leveling and PSI measurements. The stochastic model parameters of spatially and non-spatially correlated deformation regimes has been estimated using variance component estimation within the SuRe methodology (Houtenbos, 2004), see Sect. 2.3.4. The subsidence prognosis for the observation period 1992–1999 is depicted in Fig. 6.58. Table 6.6 lists the estimated variance components and their precision for three runs: one for leveling and two for InSAR with varying temporal sampling. Figures 6.59 and 6.60 show the estimated spatially correlated deformation signal and its precision respectively.

From Fig. 6.59 and Fig. 6.60 it can be deduced that also a contaminated signal of interest can be precisely estimated ($\sigma \sim 2$ mm), when taking the displacements due to benchmark settling and subsidence prognosis imperfections into account in the stochastic model, together with variance component estimation and a sufficient

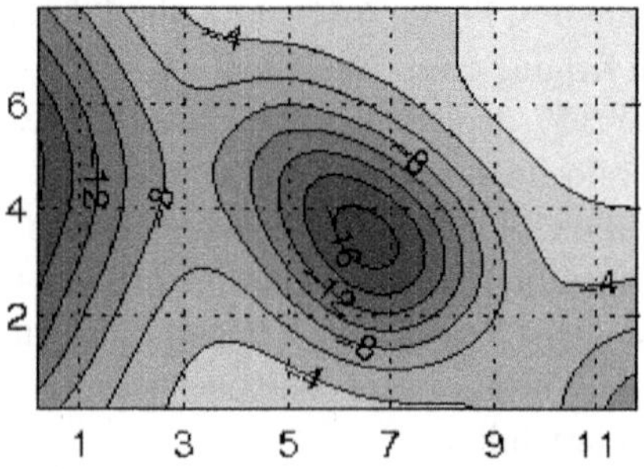

Fig. 6.58 Subsidence prognosis above the Rotterdam field (distance in km, surface displacements in mm)

Table 6.6 Estimated variance components and their precision in the estimation of the subsidence signal above the Rotterdam gas field

Type	Leveling	InSAR	InSAR
Observations	716	1698	2830
Unknowns	244	567	567
σ_{obs} (mm)	0.83±0.03	0.89±0.04	0.60±0.01
σ_{stb} (mm)	0.70±0.05	0.89±0.06	0.96±0.03
σ_{mod} (mm)	0.93±0.13	0.68±0.13	1.25±0.13
L (m)	1933±335	1887±237	2479±240
p	0.89±0.02	0.98±0.02	0.96±0.01

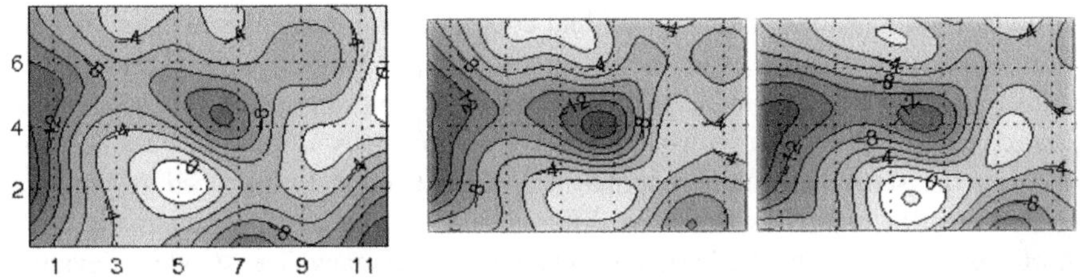

Fig. 6.59 Estimated spatially correlated deformation (mm) between 1992 and 1999 using (left) leveling, (middle) PSI and (right) PSI with a higher temporal sampling

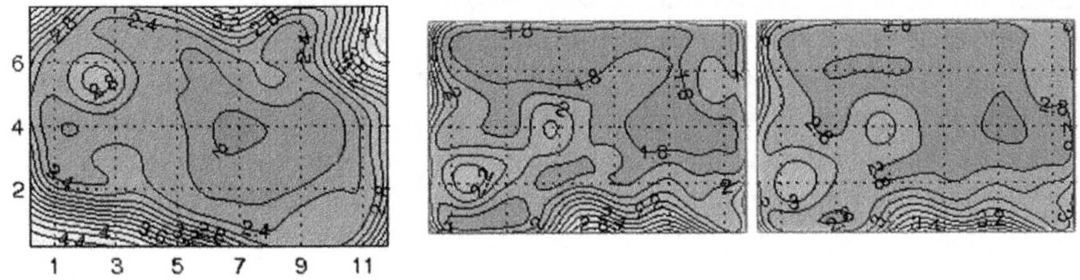

Fig. 6.60 Precision (1-sigma, mm) of the displacement component that is spatially correlated, for (left) leveling, (middle) PSI and (right) PSI with a higher temporal sampling

spatial and temporal measurement density. Leveling and PSI generally show the same subsidence pattern. This subsidence pattern is the accumulated subsidence due to all spatially correlated deformation regimes.

It can be seen that the estimated deformation patterns from both leveling and PSI deviate from the subsidence prognosis. In this particular case study, the spatially correlated deviations from the subsidence prognosis could be explained by physically ongoing processes in the subsurface: water injection in this period has caused relatively less subsidence than predicted in the prognosis.

6.6 Conclusions

This chapter has shown that a coherent deformation signal can be estimated using PSI, even in rural areas that are affected by temporal decorrelation and are subject to low subsidence rates (<1 cm/year) over a wide spatial extent. The subsiding areas correspond with the location of the gas fields that are in production. The quality description of the deformation estimates has been viewed from the perspective of precision and reliability of the PSI technique itself and the idealization precision for the estimation of the deformation signal of interest.

The PS density in Groningen follows buildings and other man-made features in the terrain, and varies from 0–10 PS/km^2 in rural areas up to more than 100 PS/km^2 in urban areas. The precision of the displacement rates is $\sim$0.1–0.5 mm/year both for ERS and Envisat. The precision of the displacement estimates is $\leq$3 mm in urban areas, and 3–7 mm in rural areas. The interferometric phase observations from Envisat contain systematic residual components that have been estimated and removed prior to the PSI estimation.

A reliability assessment has been performed utilizing six overlapping ERS tracks that cover the Groningen subsidence bowl. Redundancy is introduced since they monitor the same deformation signal. The datum connection procedure has integrated the PSI estimates from the different tracks, that cover together the entire northeastern part of the Netherlands and a part of Germany. The standard deviation of the PS velocities is less than 1 mm/year after datum connection for 70% of the multi-track PS clusters. Moreover, the deformation was decomposed into one horizontal and the vertical component. The magnitude of the horizontal components (2–3 mm/year) approximately agrees with the theoretically expected horizontal movements for the Groningen subsidence bowl (maximal 3 mm/year).

The idealization precision of PSI for monitoring subsidence due to gas extraction can be improved by PS characterization and the use of a-priori knowledge on the spatio-temporal behavior of the deformation signal. PS characterization is based on the assumption that phase observations from direct reflections that refer to well-founded buildings are most representative for subsidence due hydrocarbon production. PS characterization methods (Alternating Polarization, PS heights, and reflectivity as a function of viewing geometry) have been evaluated based on the PS velocity histograms for two case study areas. In these case study areas, PS selection based on characterization parameters resulted in a shift towards PS velocities of a lower magnitude, but this shift appeared to be not significant ($<$0.5 mm/year). Hence, in these case study areas, it suffices to apply spatial correlation of subsidence due to gas extraction for PS selection to increase the idealization precision for deformation monitoring.

Chapter 7
Cross-Validation and Operational Implementation

In Chap. 6, the accuracy (precision and reliability) of the PSI technique has been evaluated. The precision of a displacement estimate varies from ≤ 3 mm in urban areas to 3–7 mm in rural areas, see Sect. 6.3.1. The precision of displacement rates is $\sim$0.1–0.5 mm/year. The reliability of the PSI results has been demonstrated in the multi-track datum connection procedure in Sect. 6.4. Moreover, it has been shown that the idealization precision of PSI for deformation monitoring can be improved by PS characterization and PS selection based on spatial correlation of the signal of interest, see Sect. 6.5.

For the acceptance of PSI as an operational deformation monitoring technique, it has to be demonstrated that the deformation estimates are free of unquantified systematic effects. Because the majority of the gas fields in the Netherlands were already in production before the radar satellites were operational, the agreement with subsidence estimated from historical measurements is essential. Therefore, this chapter addresses the comparison of PSI deformation estimates with leveling measurements. The comparison is performed from an integrated perspective: the uncertainties of both techniques are considered and they should agree within the error margins.

Since leveling and PSI have complementary characteristics (high precision observations versus high spatial density), the spatial and temporal observation density and the precision that can be obtained play an important role. This is addressed from a theoretical perspective in Sect. 7.1. Section 7.2 contains the comparison of leveling and PSI deformation estimates in the period from 1993 until 2007. To conclude, Sect. 7.3 proposes a mathematical framework for the integration of geodetic observations of multiple techniques in the estimation of a common deformation signal.

7.1 Precision and Spatio-Temporal Observation Frequency

This section compares PSI and leveling in terms of precision and spatio-temporal observation frequency. Although PSI and leveling are both used for deformation monitoring, the type of observations is not the same. This is addressed in the first part of this section. Subsequently, the performance of both techniques is evaluated based on precision measures of the deformation estimates.

7.1.1 PSI and Leveling Deformation Estimates

The difference between leveling and PSI observations is depicted in Fig. 7.1. Leveling heights are estimated from spatial height difference measurements. From subse-

V.B.H. (Gini) Ketelaar, *Satellite Radar Interferometry*,
Remote Sensing and Digital Image Processing 14,

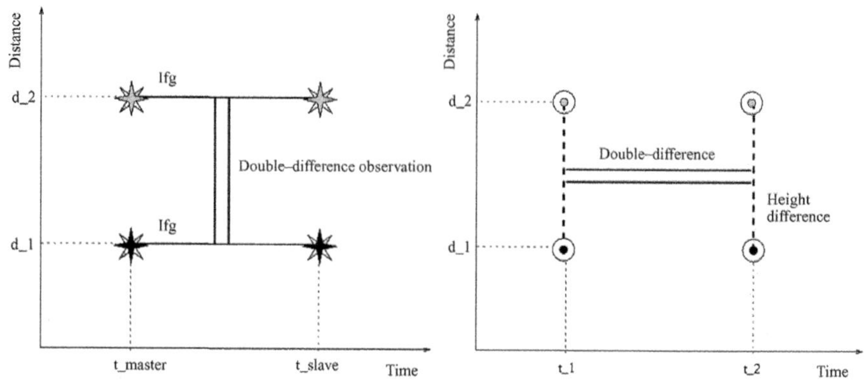

Fig. 7.1 PSI and leveling double-differences: a PSI observation is an interferometric phase difference in time; the double-difference is formed by a spatial difference between two PS (left). A leveling observation is a spatial height difference; the double-difference is formed by a temporal difference between two epochs (right)

quent leveling epochs, leveling benchmarks displacements are deduced. The basic PSI observation is an interferometric phase difference between two epochs. The first information-bearing observation, the double-difference, is formed between two PS at different locations.

Furthermore, leveling and PSI observations differ in the spatial projection of the measurements. PSI double-differences are measured along the satellite line of sight, whereas leveling heights are measured in a vertical datum, see Sect. 4.4.3.

Moreover, PSI and leveling displacements are physically not the same: orthometric and ellipsoidal displacements are distinguished. Orthometric displacements are displacements that are relative to the geoid, which is the equipotential surface of the earth's gravity field that follows the global mean sea level. Leveling heights are orthometric heights. On the other hand, PSI deformation estimates are ellipsoidal displacements. They refer to a mathematical ellipsoidal shape of the earth. Also GPS heights are ellipsoidal.

The difference between ellipsoid and geoid can be determined from gravity measurements and results in a geoid model, e.g., EGM96. The difference between ellipsoidal heights H and orthometric heights h is equal to the geoid height N (geoid undulation):

$$N = H - h. \tag{7.1}$$

The difference in geoid height in Groningen is in the order of 50 centimeters over a distance of several tens of kilometers. The relative precision of the geoid between two points i and j over a distance d is (RDNAP, 2008):

$$\sigma_{N_{ij}}\ (\text{cm}) = 0.35 + 0.003d\ (\text{km}) \tag{7.2}$$

which is less than a centimeter over a 200 kilometers distance.

Neglecting the geoid in the integration of geometric and orthometric displacements will result in a systematic effect over a large extent. However, when using double-differences as basic observations, the difference between geoid and ellipsoid cancels, assuming that the geoid does not change in time.

7.1.2 Setup for the Evaluation of Spatio-Temporal Sampling

For deformation analysis, both leveling and PSI deformation estimates are evaluated in time and space. Both techniques are relative techniques. To compare them, deformation estimates from both leveling and PSI are defined as double-difference displacements:

$$\underline{d}_{ij}^{t_1t_2}, \tag{7.3}$$

relative in time (between t_1 and t_2) and space (between points i and j).

To evaluate the performance for deformation monitoring of both PSI and leveling, the following two strategies are proposed:

1. comparison of double-difference displacements along the vertical: derivation of double-difference displacements from the leveling height estimates and projection of PSI displacement estimates along the vertical,
2. evaluation of the precision of deformation estimates using the PSI system of equations: leveling height estimates are converted to double-difference observations along the satellite line of sight.

The first option obtains the same projection of the leveling and PSI displacements by converting the PSI displacement estimates from satellite line of sight to the vertical. For leveling, the double-difference displacements are constructed from the height estimates $\hat{\underline{h}}$ per epoch:

$$\underline{d}_{ij}^{t_1t_2} = \hat{\underline{h}}_j^{t_2} - \hat{\underline{h}}_i^{t_2} - \hat{\underline{h}}_j^{t_1} + \hat{\underline{h}}_i^{t_1}. \tag{7.4}$$

The second option is based on the PSI system of equations. Because the focus lies on the spatial and temporal sampling, it is assumed that all PSI phase observations have been unwrapped correctly. The interferometric phase observations are defined between master time $t_1 = t_m$ and slave time $t_2 = t_s$. Their functional relation with the double-difference displacements along the vertical reads:

$$\underline{\varphi}_{ij}^{t_mt_s} = -\frac{4\pi\cos\theta_i^m}{\lambda}\underline{d}_{ij}^{t_mt_s}, \tag{7.5}$$

where θ_i^m is the incidence angle, and double-difference displacements are defined as:

$$d_{ij}^{t_mt_s} = d_{ij}^{t_s} - d_{ij}^{t_m}, \tag{7.6}$$

while double-differences phase observations are defined as:

$$\varphi_{ij}^{t_mt_s} = \varphi_{ij}^{t_m} - \varphi_{ij}^{t_s}. \tag{7.7}$$

Redundancy is introduced in the PSI system of equations by modeling the deformation between two PS as a linear displacement rate under the null hypothesis. Restricting the unknowns to the linear displacement rates v and the (residual) topographic heights H, the function model reads (see (6.4)):

$$\underline{\varphi}_{ij}^{k} = -\frac{4\pi}{\lambda} T^k v_{ij} - \frac{4\pi}{\lambda} \frac{B_i^{\perp}}{R_i^m \sin\theta_i^m} H_{ij} + \underline{e}, \tag{7.8}$$

where $\underline{e}$ represents measurement noise, deformation model imperfections, and (residual) atmospheric signal, and k represents the kth interferometric combination in a single master stack. The corresponding equation for leveling differs in the following ways:

- the absence of the (residual) topographic height unknowns,
- the absence of stochastically modeled atmospheric disturbances.

The observations are constructed from the leveling heights in the following way:

$$\underline{\varphi}_{ij}^{t_1 t_2} = -\frac{4\pi \cos\theta_i^m}{\lambda} (\underline{h}_{ij}^{t_2} - \underline{h}_{ij}^{t_1}). \tag{7.9}$$

If prognosis grids z of the deformation signal are available, they can be subtracted from the double-difference phase observations after conversion to phases along the satellite line of sight:

$$\underline{\varphi}_{ij}^{k} + \frac{4\pi \cos\theta_i^m}{\lambda} z_{ij}^k = -\frac{4\pi}{\lambda} T^k v_{ij} - \frac{4\pi}{\lambda} \frac{B_i^{\perp}}{R_i^m \sin\theta_i^m} H_{ij} + \underline{e}. \tag{7.10}$$

To provide insight in the effect of spatial and temporal sampling on the precision of deformation estimates and the unknown stochastic model parameters, the redundant PSI system of equations and the corresponding variant for leveling have been utilized. It has been assumed that the PSI double-differences have been unwrapped with a success rate of 1, and that the subsidence rates are linear.

For PSI, (a subsample of) the configuration of the six Groningen tracks has been utilized (temporal baselines, perpendicular baselines, Doppler centroid frequencies). A series of subsidence prognoses based on geomechanical subsurface modeling has been available (1964–2007, eight predictions with intervals of ~5 years). The precision and the spatio-temporal observation frequency of the PSI and leveling technique have been evaluated based on the precision of the deformation estimates and the precision of the stochastic model parameters.

7.1.3 Temporal Sampling

This section addresses the effect of temporal sampling on the precision of the deformation estimates from PSI and leveling. Simulations of leveling and PSI observations have been performed, based on the assumption that benchmark and PS

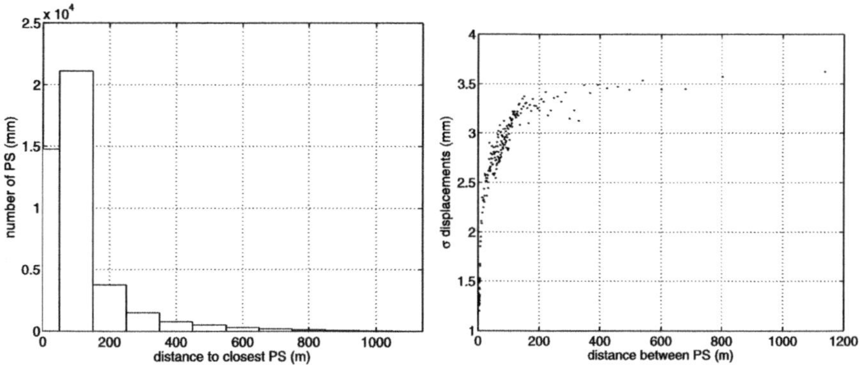

Fig. 7.2 Left: histogram of mutual PS distances (distance to the closest PS). Right: standard deviation of the Envisat PSI displacements as a function of distance between the PS, after outlier removal based on spatial correlation. Although the majority of the PS displacements have a standard deviation of ~3 mm, the standard deviation of two PS at a one kilometer distance is slightly higher

displacements are linear in time. The mathematical model that is described by (7.8) has been used for both PSI and leveling. The precision of the estimated displacement rates are evaluated based on the variance-covariance matrix. The quality of the variance-covariance matrix is parameterized by the Dilution of Precision for PSI (DOP_{PSI}). This precision measure has been described in Sect. 4.3.4 and is independent of the choice of the spatial and temporal reference.

For the evaluation of the temporal sampling, the standard deviation of PSI double-difference observations has been set to 3 mm (see Sect. 6.3.1), whereas the standard deviation of a leveling height difference has been set to 1 mm. It has to be noted however, that the precision of both the leveling and the PSI observations is a function of distance. The standard deviation of leveling height measurements is ~ 1 mm/$\sqrt{\text{km}}$ (de Bruijne et al., 2005). Figure 7.2 shows the standard deviation of PSI displacements between two neighboring PS as a function of the distance between two PS, for the Envisat PSI deformation estimates in the period 2003–2007, see Sect. 6.2.2. It can be deduced that the precision of relative PS displacements at a distance of one kilometer is slightly higher than 3 mm. This is partly caused by the choices that were made for the separation of atmospheric signal and unmodeled deformation, see Sect. 6.1.3. Since the emphasis in the PSI processing has been set to the preservation of unmodeled deformation, the displacement time series is slightly noisier, as it may contain residual atmospheric signal that has not been removed. It is expected that the standard deviation of the displacements will decrease with future improvements on the stochastic modeling of atmospheric signal (Hanssen, 2001; Grebenitcharsky and Hanssen, 2005; Liu et al., 2008).

Regarding redundancy and network design, the ratio between the number of height difference observations and measurement points in the leveling network has been set to 6:5. For PSI, this ratio is ~ 1, since only $P-1$ spatial differences can be constructed from P PS.

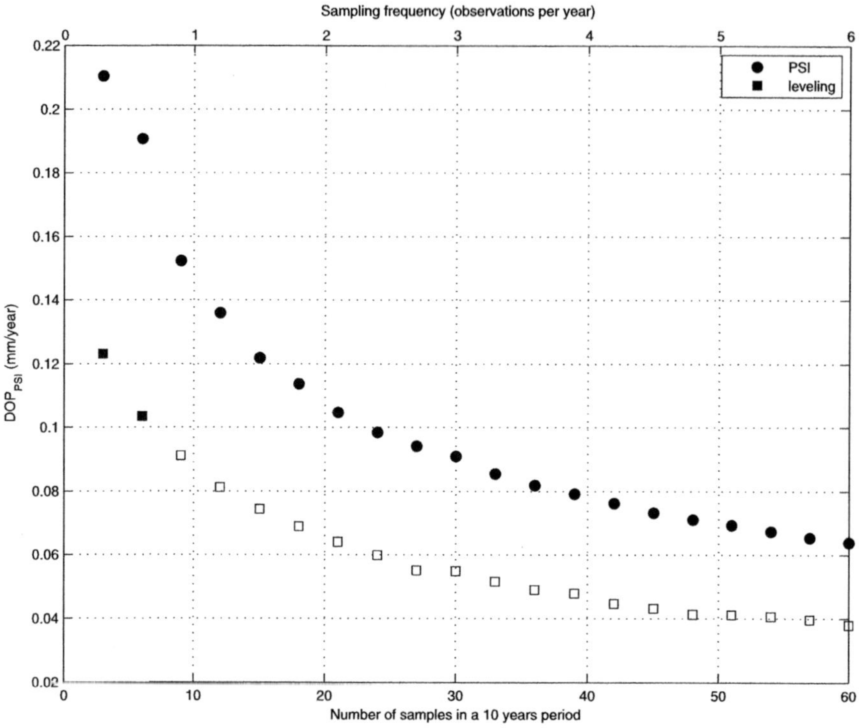

Fig. 7.3 $\mathrm{DOP}_{\mathrm{PSI}}$ values for displacement rates (velocities) as a function of temporal sampling for PSI (circles) and leveling (squares). Although PSI observations have a lower precision, the higher temporal sampling has the potential to result in a similar or even better precision of the displacement rate estimates. For leveling campaigns each 2–5 years, approximately 25 satellite acquisitions are required to obtain the same precision of displacement rates

Figure 7.3 shows the $\mathrm{DOP}_{\mathrm{PSI}}$ values for the displacement rate estimates for both PSI and leveling. Due to the lower observational precision of PSI, a time series of $\sim$25 images in 10 years is required to result in a similar precision compared to leveling campaigns each 2–5 years. In fact, for stacks that consist of more than 25 images, the temporal sampling of PSI can result in a even higher precision compared to levelings each 5 years, even though the observational precision is lower.

It has to be noted that although the DOP precision measure is independent of the spatio-temporal reference, it is one specific function of the variance-covariance matrix. The DOP measure is a scalar precision measure that is easy to compare. Nevertheless, a complete analysis of the precision represented by two variance-covariance matrices is performed through the generalized eigenvalue problem, see e.g., Teunissen et al. (2005):

$$\det(Q_{\mathrm{PSI}} - \lambda Q_{\mathrm{lev}}) = 0, \tag{7.11}$$

where Q_{PSI} and Q_{lev} are the variance-covariance matrices of the estimated displacement rates for PSI and leveling respectively. Eigenvalues less than one indicate that the precision of PSI is better than the precision of leveling, while eigenvalues larger than one indicate that the precision of leveling is better. The best and worst precision are indicated by the smallest and largest eigenvalue respectively. The corresponding eigenvector represents the direction in the parameter space in which the precision corresponding with a certain eigenvalue is obtained.

The higher temporal sampling of InSAR has a significant advantage: model deviations in time can be detected, provided that the temporal sampling frequency exceeds twice the highest frequency of this model deviation (the Nyquist rate). This implies that with 6 acquisitions per year, model deviations over periods larger than 4 months can be detected. With an acquisition each 35 days, model deviations with a period larger than 70 days can be detected. Leveling observations, with a frequency of 2–5 years, are not able to detect model deviations over a several months' period.

7.1.4 Spatial Sampling

The spatial sampling of leveling and PSI is evaluated in this section with the purpose to monitor subsidence due to hydrocarbon production. The average PS density in the Groningen area is ~40 PS/km^2 (see Sect. 6.2.1), which is significantly higher than the leveling benchmark density of 1–2 per km^2. In urban areas, the PS density even exceeds 100 PS/km^2.

For the parameterization of the spatial subsidence pattern, that has the approximate shape of an ellipsoidal bowl, the following options are considered:

1. average displacement estimates within a certain area, e.g., a quadtree grid cell,
2. subsidence prognosis grids, and
3. a point source model.

7.1.4.1 Average Displacement Estimates

If the area of interest can be subdivided into grid cells in which the surface displacement is assumed constant, see e.g., Fig. 6.36, average displacement estimates per grid cell can be computed. We start with the situation in which the displacement estimates within a grid cell can be considered uncorrelated, and subsequently demonstrate the effect of correlation between the displacement estimates.

If the displacements are assumed uncorrelated and of equal precision σ_d, the precision of the average displacement $\underline{\hat{d}}$ equals:

$$\sigma_{\hat{d}} = \frac{\sigma_d}{\sqrt{m}}, \tag{7.12}$$

where m is the number of displacement estimates. The relationship between parameter precision and number of observations is not linear. Figure 7.4 shows the average displacement precision as a function of the number of observations with a

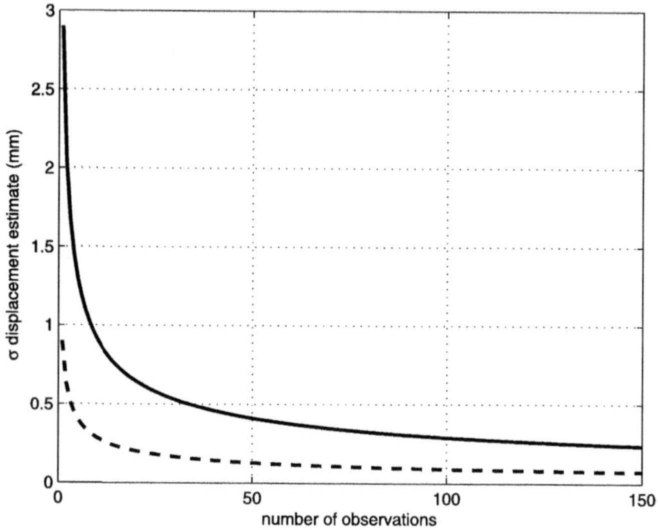

Fig. 7.4 Precision average displacement estimates as a function of number of observations for an observational precision of 1 (dashed) and 3 (solid) mm. Displacements are considered to be equal and uncorrelated within a certain spatial radius, and an average displacement is estimated. Hence the standard deviation of the average displacement equals $\sigma_d/\sqrt{m}$, where m is the number of displacements

precision of 1 and 3 mm respectively. Whether different measurement techniques can compete, depends on the required precision for the average deformation estimates. The higher the required precision of the average displacement estimates, the more additional observations are needed for the measurement technique with the lower observational precision.

The leveling benchmark density in Groningen is $\sim$1–2 per km^2. Considering m_{lev} leveling displacement estimates, and variances of $\sigma^2_{d_{\text{PSI}}}$ and $\sigma^2_{d_{\text{lev}}}$, then the required number of PSI displacements is:

$$m_{\text{PSI}} = \frac{\sigma^2_{d_{\text{PSI}}}}{\sigma^2_{d_{\text{lev}}}} m_{\text{lev}}, \tag{7.13}$$

if the PSI and leveling displacements would be uncorrelated. For a precision ratio of $\sigma_{\text{lev}} : \sigma_{\text{PSI}} = 1{:}3$, the PS density has to be 9 PS/km^2. However, in practice both PSI and leveling displacement estimates are mutually correlated. For leveling and PSI, it can be demonstrated that the precision of the average displacement estimates is dependent on respectively the network design and the correlation between the double-difference observations.

The effect of correlation on the average displacement estimates for leveling is demonstrated assuming $(P-1)$ height estimates $\underline{\hat{h}}$ from two identical measurement campaigns at time t_1 and t_2. Benchmark 1 is considered as the reference benchmark.

The height estimates and their variance-covariance matrix read:

$$\underline{\hat{x}} = \begin{bmatrix} \underline{\hat{h}}_2^{t_1} \\ \vdots \\ \underline{\hat{h}}_P^{t_1} \\ \underline{\hat{h}}_2^{t_2} \\ \vdots \\ \underline{\hat{h}}_P^{t_2} \end{bmatrix}; \qquad Q_{\hat{x}} = \begin{bmatrix} Q_{\hat{h}} & 0 \\ 0 & Q_{\hat{h}} \end{bmatrix}, \tag{7.14}$$

for uncorrelated epochs t_1 and t_2. The double-difference displacement estimates are a linear combination of the height estimates:

$$\underline{d} = \begin{bmatrix} I & -I \end{bmatrix} \begin{bmatrix} \underline{\hat{h}}_2^{t_1} \\ \vdots \\ \underline{\hat{h}}_P^{t_1} \\ \underline{\hat{h}}_2^{t_2} \\ \vdots \\ \underline{\hat{h}}_P^{t_2} \end{bmatrix}; \qquad Q_d = \begin{bmatrix} I & -I \end{bmatrix} \begin{bmatrix} Q_{\hat{h}} & 0 \\ 0 & Q_{\hat{h}} \end{bmatrix} \begin{bmatrix} I \\ -I \end{bmatrix} = 2Q_{\hat{h}}. \tag{7.15}$$

Subsequently, the mathematical model for the estimation of the average displacement reads:

$$E\left\{\begin{bmatrix} \underline{d}_2 \\ \vdots \\ \underline{d}_P \end{bmatrix}\right\} = e_m d; \qquad D\left\{\begin{bmatrix} \underline{d}_2 \\ \vdots \\ \underline{d}_P \end{bmatrix}\right\} = 2Q_{\hat{h}}. \tag{7.16}$$

Application of the propagation law of variances and covariances leads to the following precision of the average displacement estimate $\underline{\hat{d}}$:

$$\sigma_{\hat{d}}^2 = \frac{2}{\sum_{i=1}^{P-1} \sum_{j=1}^{P-1} Q_{\hat{h}}^{-1}{}_{ij}}. \tag{7.17}$$

Hence, the precision of the average displacement estimate depends on the precision of the leveling height difference observations σ_y and the leveling network that is described in the design matrix A.

In PSI, the correlation is directly introduced in the construction of double-difference phase observations, see (4.19). The relation of each double-difference phase observation with the displacement estimate is linear, see (3.11). Hence, the mutual correlation of PSI displacement estimates is similar to the mutual correlation of the double-difference phase observations. Figure 7.5 shows the effect of the correlation of double-difference phase observations on the displacement estimates. It can be seen that the precision improvement of the average displacement is overestimated if the displacements are considered uncorrelated. Moreover, the precision of

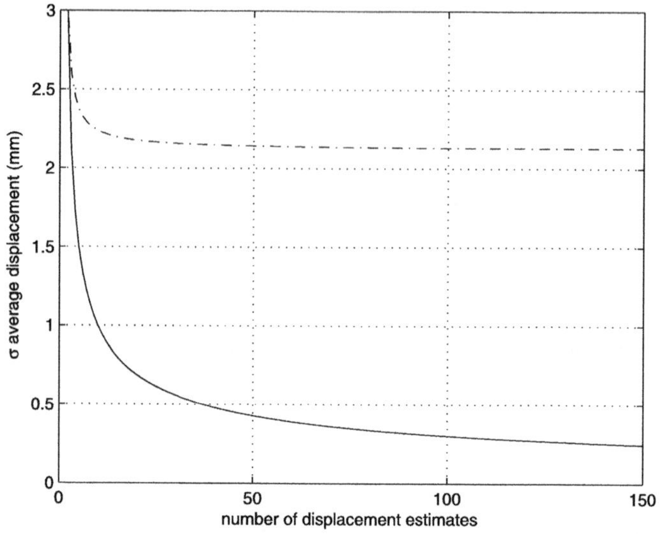

Fig. 7.5 Precision average displacement estimates as a function of number of observations for uncorrelated (solid) and correlated (dash-dot) PSI displacement estimates. Only one interferometric combination is considered; the precision of the average displacement estimates is shown for varying spatial sampling. The precision of the double-difference phase observations is 3 mm. If the correlation between the displacement estimates is neglected, the precision of the average deformation estimates is overestimated

the average PSI displacement estimate does not drop beyond 2 mm for a precision of 3 mm for double-difference phase observations. This implies that the strength of PSI relies on both the spatial *and* the temporal sampling.

7.1.4.2 Prognosis Grids

Apart from the fact that the spatial sampling improves the precision of the deformation estimates, the minimum spatial sampling to estimate the deformation signal of interest has to be considered as well. In this section, we focus on the spatial point density that is required to estimate subsidence due to gas extraction from the Groningen gas field. Since the focus lies on the spatial pattern of the subsidence signal, the correlation between the displacement estimates is neglected in this section. For the implications of correlation, the reader is referred to the previous section.

Subsidence due to gas extraction is predicted based on the geomechanical modeling of the subsurface including the reservoir, and is described by the subsidence prognosis, see Sect. 2.3.3.1. The shape of the subsidence prognosis of the Groningen gas field and surrounding gas fields is shown in Fig. 7.6. The main subsidence bowl has an extent of ~30 km. A series of simulations has been performed where a number of observations with a certain precision are randomly selected. Subsequently, the observations were interpolated (linear interpolation) and compared to

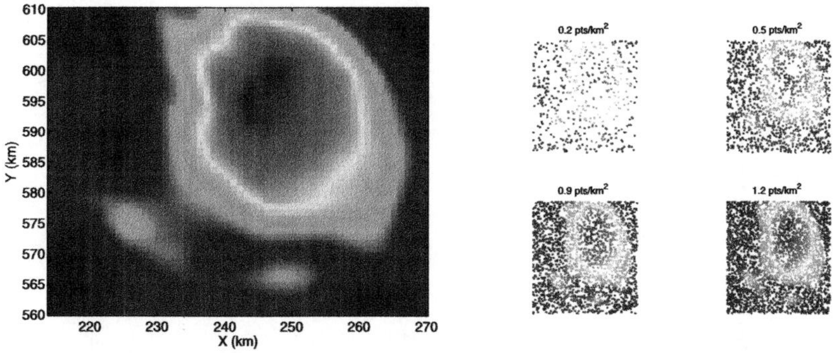

Fig. 7.6 Subsidence prognosis (500 × 500 meter grid) in the Groningen area (left). The coordinates are in the Dutch RD system. Four scenario's of spatial sampling of the subsidence prognosis are depicted (right)

the original subsidence prognosis. The match between the subsidence prognosis and the interpolated prognosis is expressed by the standard deviation of the residuals between the two. These residuals are a superposition of residual components due to the precision of the observations, the precision of the interpolator, and the spatial variation of the subsidence pattern. Figure 7.7 shows the standard deviation of the residuals between the interpolated prognosis and the subsidence prognosis as a function of the number of observations. It can be concluded that the standard deviation of the residuals decreases significantly up to a point density of 1 point per km^2. Apparently, a point density of 1 point per km^2 is sufficient to capture the spatial shape of subsidence due to gas extraction in the Groningen area.

7.1.4.3 Point Source Model

Besides the prognosis grids, subsidence due to gas extraction can be characterized by a limited number of geophysical parameters using the nucleus-of-strain concept (Geertsma, 1973a) or Mogi sources (Anderson, 1936; Mogi, 1958), see Sect. 2.3.3.1. Figure 7.8 shows an attempt to model the Groningen subsidence pattern using a single point source. The parameters that describe the single point source are:

- a multiplication factor that represents the geophysical reservoir properties (compaction coefficient, Poisson's ratio, pressure drop, and volume change),
- the source depth D, and
- the source location x_c, y_c.

From Fig. 7.8 it can be deduced that a single point source highly simplifies the actual subsidence pattern. This is confirmed by Fig. 7.9, which shows the precision of the source location and depth, that is not much affected by the spatial point density. The range of the spatial sampling in Fig. 7.9 is 0–0.5 points per km^2: this is the range in which spatial sampling significantly affects the estimation of the subsidence prognosis, see Fig. 7.7. However, the precision of the source parameters does

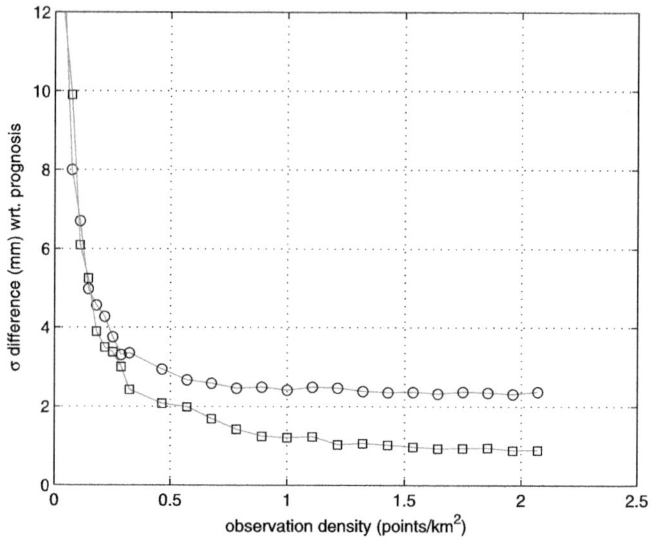

Fig. 7.7 Standard deviation of the residuals (mm) between the interpolated subsidence prognosis (linear interpolation) and the original subsidence prognosis for varying spatial point densities and for different observational precision (squares and circles correspond with 1 and 3 mm σ respectively). Up to an observation density of 1 point per km^2, the standard deviation of the residuals is a superposition of the precision of the observations, the precision of the interpolator, and the spatial variation of the subsidence pattern. A point density of 1 point per km^2 is sufficient to sample the spatial variation of the subsidence pattern

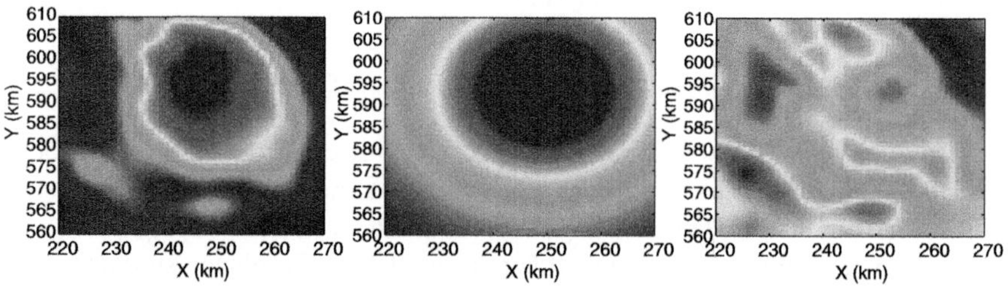

Fig. 7.8 Subsidence prognosis from start of gas extraction up to 2007 (left); subsidence prognosis modeled by a point source (middle); residual deformation signal (right). The single point source model highly simplifies the predicted subsidence pattern: the residuals cover a range of 15 cm. For comparison: the shape of the interpolated deformation signal that has been estimated from PSI is depicted in Fig. 6.36

not change significantly in this spatial sampling range. Since a single point source is not sufficient to describe subsidence due to gas extraction, we shift the focus to the spatial sampling of the residual signal.

To analyze the effect of spatial sampling on the estimation of the residual signal, realizations of residual signal with different spatial correlation lengths have been simulated. Since the depth of the gas reservoirs in the northern part of the Nether-

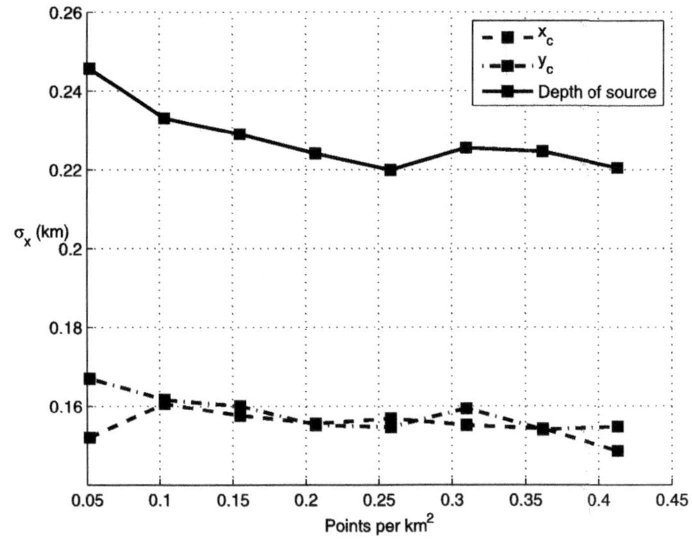

Fig. 7.9 Precision of the point source location and depth parameter estimates as a function of point density. The point density is chosen in the range 0–0.5 points per km^2: the range in which spatial sampling significantly affects the estimation of the subsidence prognosis, see Fig. 7.7. Since the single point source model highly simplifies the actual subsidence pattern, the point density does not affect the precision of the parameter estimates

lands is ~3 km, a correlation length of 3 km has been chosen. Moreover, simulations have been performed with a smaller correlation length of 1 km, which indicates the potential identification of irregularities in the subsidence pattern, e.g., due to disconnected reservoirs. In a similar way as for the prognosis grids (see Fig. 7.7), the interpolated signal has been compared to the original signal for different point densities.

Figure 7.10 shows that measurement techniques with a lower measurement precision but higher spatial point density can still achieve a similar or even higher precision level for the estimation of the residual signal. It also shows that the situation in which the difference between the original and the interpolated signal does not change significantly anymore, is reached for a relatively lower spatial sampling if the correlation length is larger. This is explained by the fact that signal with a larger correlation length requires a lower spatial sampling to fully reconstruct the signal. Furthermore, it has to be realized that a low precision measurement technique with a high point density has abilities to detect residual signal patterns that a measurement technique with a lower point density but higher observational precision would never be able to detect.

7.2 Comparison of PSI and Leveling Deformation Estimates

In Sect. 7.1 it has been shown by means of simulations that PSI can meet the precision level of the leveling technique due to its spatial and temporal sampling. In this

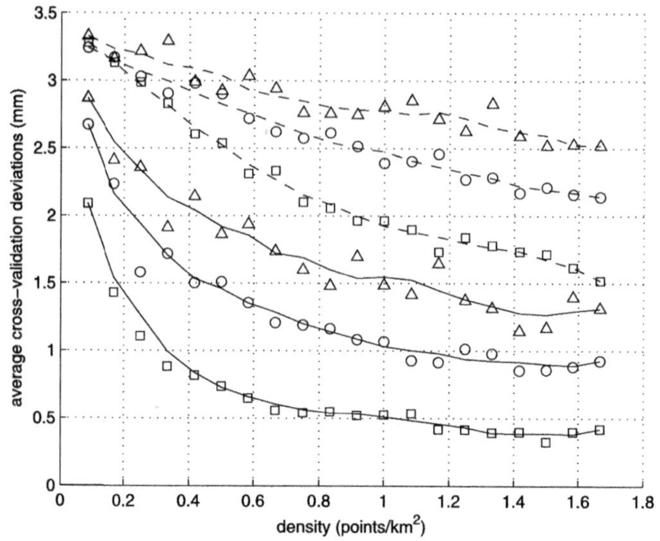

Fig. 7.10 Average difference (absolute) between the interpolated and the original residual signal, for different spatial correlation lengths of the residual signal. Solid lines correspond with a relatively smooth signal (correlation length of 3 km), while dashed lines represent a more rough signal (correlation length 1 km). Moreover, the residual signal has been simulated using three different measurement precisions: 1 mm (squares), 3 mm (circles), and 5 mm (triangles). A signal with a larger correlation length requires a lower spatial sampling density. Hence, the stage in which the difference between the original and the interpolated signal does not change significantly anymore, is reached at a lower spatial sampling frequency. Furthermore, it is shown that a measurement technique with a lower measurement precision, but higher spatial density can potentially achieve a similar or even higher precision level for the estimation of the residual signal

section, the actual PSI and leveling deformation estimates in the period from 1993 to 2007 are compared. For the comparison of leveling and PSI, the PSI deformation estimates are converted to metric units along the vertical, as explained in Sect. 7.1.1. Two basic deformation estimates are distinguished:

1. the average displacement rates (velocities) in mm/year, and
2. displacements per epoch in mm.

Both the velocities and the displacements are double-differences. Hence, the difference between orthometric and ellipsoidal estimates cancels, provided that the geoid does not change in time.

This section evaluates the similarity between leveling and PSI velocities and displacement estimates by means of the correlation coefficient and geodetic teststatistics. In the comparison, the uncertainties of both techniques are taken into account. A statement is given on the agreement between both techniques and possible explanations for deviations are addressed.

7.2.1 Parameterization of the Agreement between PSI and Leveling

The evaluation of the agreement between PSI and leveling displacement estimates requires the parameterization of the match between both techniques. In this section, we address subsequently the correlation coefficient, point-wise teststatistics, and the overall model test (OMT).

7.2.1.1 Correlation Coefficient

The correlation coefficient ρ is the standardized covariance between two variables $\underline{x}_1$ and $\underline{x}_2$, see e.g., Chatfield (1989) and Teunissen et al. (2005). It is a measure of the strength of the linear relationship between two variables. Its definition reads:

$$\rho(\underline{x}_1, \underline{x}_2) = \frac{C(\underline{x}_1, \underline{x}_2)}{\sigma_{x_1}\sigma_{x_2}} = \frac{E\{(\underline{x}_1 - E\{\underline{x}_1\})(\underline{x}_2 - E\{\underline{x}_2\})\}}{\sigma_{x_1}\sigma_{x_2}}, \tag{7.18}$$

where C indicates the covariance between $\underline{x}_1$ and $\underline{x}_2$. Values of the correlation coefficient are in the range from -1 to $+1$. A correlation coefficient of 0 indicates that the variables are uncorrelated.

The correlation coefficient is invariant for a linear transformation (scaling and offset) of the variables $\underline{x}_1$ and $\underline{x}_2$. For example, for a linear transformation of $\underline{x}_1$ to $\underline{x}'_1 = a \cdot \underline{x}_1 + b$, the offset b cancels in the computation of $\underline{x}'_1 - E\{\underline{x}'_1\}$:

$$\underline{x}'_1 - E\{\underline{x}'_1\} = a \cdot \underline{x}_1 + b - a \cdot E\{\underline{x}_1\} - b = a(\underline{x}_1 - E\{\underline{x}_1\}). \tag{7.19}$$

Moreover, from $\sigma_{x'_1} = a \cdot \sigma_{x_1}$ follows that the scaling factor a cancels in the standardization of the covariance with the standard deviation of the variables in (7.18).

The invariance of the correlation coefficient for a linear transformation implies that the correlation coefficient as a measure for the agreement between PSI and leveling displacements is not sensitive for an offset or scaling factor in the displacements. Since both PSI and leveling are relative techniques and may refer to a different reference point, the invariance for an offset in the displacements does not affect the suitability of the correlation coefficient as a measure for the match between PSI and leveling. However, the invariance for scaling is not desirable. A scaling effect would introduce a systematic underestimation or overestimation of subsidence due to gas extraction, compared to the stable areas. Nevertheless, a scaling effect has not been detected in the datum connection procedure of the six overlapping ERS tracks in the Groningen area, see Sect. 6.4. A direct physical cause for a scaling effect in the PSI displacements cannot be indicated, unless when using an erroneous value for the wavelength. Hence, the correlation coefficient can be used as a measure for the agreement between PSI and leveling displacements.

There exists a relation between the correlation coefficient and geodetic teststatistics (Teunissen, 2000b). To demonstrate this relationship, we consider the vectors of displacement estimates $\hat{\underline{x}}_1$ and $\hat{\underline{x}}_2$ for PSI and leveling respectively. The teststatistic for the match between PSI and leveling displacement estimates is defined under the null hypothesis that states that the misclosures between the displacement estimates of both techniques are zero:

$$H_0: \quad B^T E\{\underline{y}\} = [I \; -I]\, E\left\{\begin{bmatrix}\hat{\underline{x}}_1 \\ \hat{\underline{x}}_2\end{bmatrix}\right\} = E\{\hat{\underline{x}}_1 - \hat{\underline{x}}_2\} = 0. \tag{7.20}$$

The corresponding teststatistic T is the quadratic form of the misclosures in the metric of the variance-covariance matrix of the displacement estimates:

$$T = (\hat{\underline{x}}_1 - \hat{\underline{x}}_2)^T (Q_{\hat{x}_1} + Q_{\hat{x}_2})^{-1} (\hat{\underline{x}}_1 - \hat{\underline{x}}_2), \tag{7.21}$$

where the displacement estimates of both measurement techniques are considered to be uncorrelated. The null hypothesis is rejected if the value of T is larger than the critical value k_α, where α is the size of the type I error (rejecting H_0 while in fact H_0 is true). The relation of the teststatistic of (7.21) with the correlation coefficient can be seen if we consider the quadratic form $(\hat{\underline{x}}_1 - \hat{\underline{x}}_2)^T(\hat{\underline{x}}_1 - \hat{\underline{x}}_2)$, under the assumption that both $Q_{\hat{x}_1}$ and $Q_{\hat{x}_2}$ are equal to the identity matrix I:

$$\begin{aligned}\hat{\underline{x}}_1^T\hat{\underline{x}}_1 - 2\hat{\underline{x}}_1^T\hat{\underline{x}}_2 + \hat{\underline{x}}_2^T\hat{\underline{x}}_2 &= (\|\hat{\underline{x}}_1\| - \|\hat{\underline{x}}_2\|)^2 + 2\|\hat{\underline{x}}_1\|\|\hat{\underline{x}}_2\| - 2\hat{\underline{x}}_1^T\hat{\underline{x}}_2 \\ &= (\|\hat{\underline{x}}_1\| - \|\hat{\underline{x}}_2\|)^2 + 2\|\hat{\underline{x}}_1\|\|\hat{\underline{x}}_2\|\left(1 - \frac{\hat{\underline{x}}_1^T\hat{\underline{x}}_2}{\|\hat{\underline{x}}_1\|\|\hat{\underline{x}}_2\|}\right),\end{aligned} \tag{7.22}$$

where the last term is equal to the estimator of the correlation coefficient if the expectation of both variables equals 0:

$$\frac{\hat{\underline{x}}_1^T\hat{\underline{x}}_2}{\|\hat{\underline{x}}_1\|\|\hat{\underline{x}}_2\|} = \cos\beta, \tag{7.23}$$

where β is the angle between the displacement vectors of $\hat{\underline{x}}_1$ and $\hat{\underline{x}}_2$.

The teststatistic as defined in (7.21) is invariant for a linear transformation of both $\hat{\underline{x}}_1$ and $\hat{\underline{x}}_2$. A linear transformation with matrix U and offset vector v

$$\hat{\underline{x}}_1' = U\hat{\underline{x}}_1 + v; \qquad \hat{\underline{x}}_2' = U\hat{\underline{x}}_2 + v \tag{7.24}$$

would lead to the teststatistic:

$$T = (\hat{\underline{x}}_1 - \hat{\underline{x}}_2)^T U^T (U(Q_{\hat{x}_1} + Q_{\hat{x}_2})U^T)^{-1} U(\hat{\underline{x}}_1 - \hat{\underline{x}}_2), \tag{7.25}$$

which is equal to the teststatistic in (7.21). However, besides the properties of the teststatistic itself, we focus on the effect of a transformation of $\hat{\underline{x}}_1$ with respect to $\hat{\underline{x}}_2$. This transformation quantifies the agreement between deformation estimates obtained from InSAR and leveling. It has been shown that the correlation coefficient, see (7.18), is invariant for scaling or an offset in $\hat{\underline{x}}_1$ with respect to $\hat{\underline{x}}_2$. However, the teststatistic from (7.22) does quantify the relative difference between $\hat{\underline{x}}_1$ and $\hat{\underline{x}}_2$. Both an offset and a scaling factor will affect the value of the teststatistic. Although the correlation coefficient is an easily interpretable value that can be applied to quantify the match between PSI and leveling, geodetic teststatistics will also take respective scaling effects and offsets between the two techniques into account.

7.2.1.2 Point-wise Teststatistics

Point-wise teststatistics are based on the misclosures between the deformation estimates $\underline{\hat{x}}_1$ and $\underline{\hat{x}}_2$ from PSI and leveling respectively for a single evaluation location (PS or benchmark). A model of condition equations can be constructed, where estimates from $\underline{\hat{x}}_1$ and $\underline{\hat{x}}_2$ for location i are the input in the vector of observations:

$$B^T E\{\underline{y}\} = [1\ -1]\, E\left\{\begin{bmatrix}\underline{\hat{x}}_{1(i)}\\ \underline{\hat{x}}_{2(i)}\end{bmatrix}\right\} = 0; \qquad Q_y = \begin{bmatrix}\sigma^2_{\hat{x}_{1(i)}} & 0\\ 0 & \sigma^2_{\hat{x}_{2(i)}}\end{bmatrix}. \tag{7.26}$$

The corresponding teststatistic and its theoretical distribution read:

$$\underline{T}_{q=1} = \underline{t}^T Q_t^{-1} \underline{t} \quad \sim \chi^2(1,0), \tag{7.27}$$

where $\underline{t}$ is the misclosure between the PSI and leveling deformation estimates at location i.

The point-wise teststatistics from (7.26) have the disadvantage that they do not take into account the mutual correlation between the displacement estimates from both PSI and leveling. Therefore, a general set up is now defined.

7.2.1.3 Overall Model Test

PSI and leveling displacement estimates are uncorrelated. However, *within* each technique, all displacement estimates are correlated due the network construction and the double-differences. To take into account these correlations, a general model of condition equations is set up in the following way:

$$B^T E\{\underline{y}\} = [I\ -I]\, E\left\{\begin{bmatrix}\underline{\hat{x}}_1\\ \underline{\hat{x}}_2\end{bmatrix}\right\} = 0; \qquad Q_y = \begin{bmatrix}Q_{\hat{x}_1} & 0\\ 0 & Q_{\hat{x}_2}\end{bmatrix}, \tag{7.28}$$

which is equal to (7.20). The corresponding teststatistic for the overall model test with its theoretical distribution reads:

$$\underline{\hat{\sigma}}^2 = \frac{\underline{T}_{q=m-n}}{m-n} = \frac{\underline{t}^T Q_t^{-1} \underline{t}}{m-n} \quad \sim F(m-n, \infty, 0), \tag{7.29}$$

where $m-n$ is equal to the number of conditions, and $\underline{t}$ is the vector of misclosures between PSI and leveling deformation estimates. The null hypothesis is rejected if the overall model test exceeds the critical value k_α. Not only the overall model test can be computed using the mathematical model from (7.28), but also single point teststatistics, such as the $\underline{w}$-teststatistics for datasnooping, see Sect. 2.3.1.

7.2.2 *PSI and Leveling Displacement Rates*

The comparison and integration of PSI and leveling displacement rates can be performed in two ways, i.e., comparison at benchmark or PS level, and comparison of the interpolated subsidence signal. The disadvantage of both options find their origin in the possible existence of multiple deformation regimes, see Sect. 4.5.1. The first option has the disadvantage that neighboring PS and leveling benchmarks do not necessarily represent the same deformation regime. The second option can lead to an inaccurate interpolation because of uncertainties in the stochastic model parameters that describe the spatio-temporal behavior of the deformation regimes. In this section, we choose to perform the comparison of the deformation estimates at benchmark level (point-wise). In this way, outliers can be attributed to specific benchmarks and PS.

7.2.2.1 Comparison of PSI and Leveling Displacement Rates

Point-wise comparison of PSI and leveling displacement rates is performed at the leveling benchmark locations, in the period that is covered by both leveling and PSI: 1993–2003. For leveling, the displacement estimates from the subsidence analysis over the period 1964–2003 have been utilized (Schoustra, 2004). The Subsidence Modeling procedure (Odijk and Kenselaar, 2003) has been followed in this analysis, that has been applied to a subset of benchmarks that are considered to represent subsidence due to hydrocarbon production ('stable benchmarks'), see Sect. 6.5.2. The leveling displacements have been converted to displacement rates for the comparison with the PS velocities.

Since datasnooping has been performed on the leveling observations and moreover stable benchmarks have been selected, outliers have been removed from the PS results as well. Due to the data dimensions, PS have been grouped into grid cells of 5×5 km, where a constant PS velocity is assumed. The difference in subsidence rates within a grid cell has been taken into account in the critical value in the datasnooping procedure. The maximum deviation of displacement rates within a grid cell of 5×5 km around their mean has been determined using the subsidence prognoses in the period 1993–2003, and is 2 mm/year. Adding the uncertainty in the velocity estimates (see Sect. 4.2), all PS have been removed that exhibit a velocity difference of more than 3 mm/year with respect to the mean velocity in the grid cell. This coarse datasnooping procedure has been followed by outlier removal based on the quadtree decomposition of the deformation signal estimated from PSI. Additionally, one constant offset has been applied to all PS velocities, based on the assumption that the majority of the covered area is not affected by deformation and can be considered stable (0 mm/year).

The construction of displacement rates does not necessarily imply that subsidence is linear in time. In the majority of the gas fields in the northern part of the Netherlands the movements are (near) linear in the period from 1993, but some of the smaller fields have been taken into production in a later stage. However, PSI and

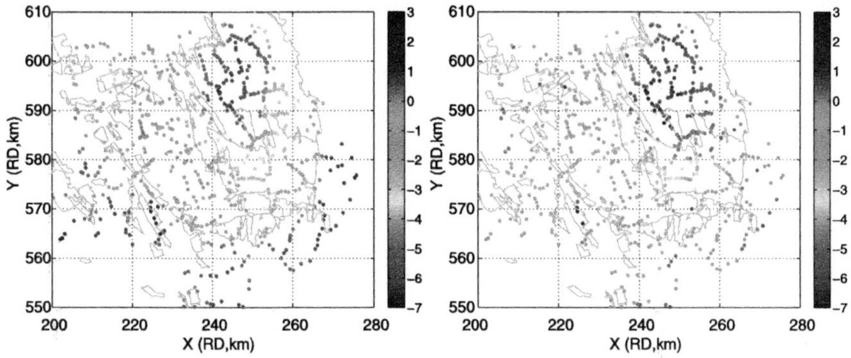

Fig. 7.11 Leveling displacement rates per benchmark (mm/yr) and the average displacement rates of the PS within a 500 meter radius around the benchmarks. For PSI, gross outliers have been removed by assuming spatial correlation of the displacement rates. Subsequently, the mean of the PS displacement rates at each benchmark location has been computed

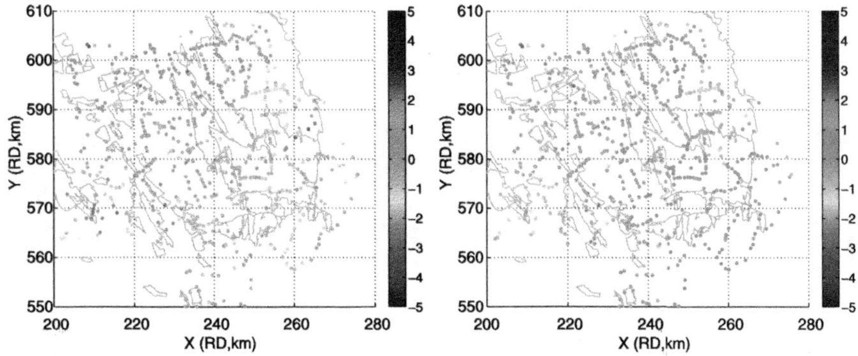

Fig. 7.12 Differences between PSI and leveling displacement rates: PSI—leveling (mm/year). Both before (left) and after correction of a bias and spatial trend (right), areas can be identified in which the PSI displacement rates are slightly larger than the leveling displacement rates

leveling should result in similar displacement estimates and hence in approximately similar displacement rates, depending on the temporal sampling. For each leveling benchmark, PS have been selected within a distance of 500 m. For these selected PS, an average displacement rate has been computed. Figure 7.11 shows the displacement rates for both leveling and PSI. The spatial subsidence pattern is clearly visible for both techniques.

Figure 7.13 (left) shows the difference between the PSI and leveling displacement rates. Although the standard deviation of the differences between leveling and PSI is ~1 mm/year, it can be seen that a bias is of ~0.5 mm/year is present between the PSI and leveling displacement rates. This bias can be explained by a different spatial reference, and can be corrected for. Furthermore, it can be deduced from Fig. 7.12 that the differences between the leveling and PSI displacement rates are spatially correlated.

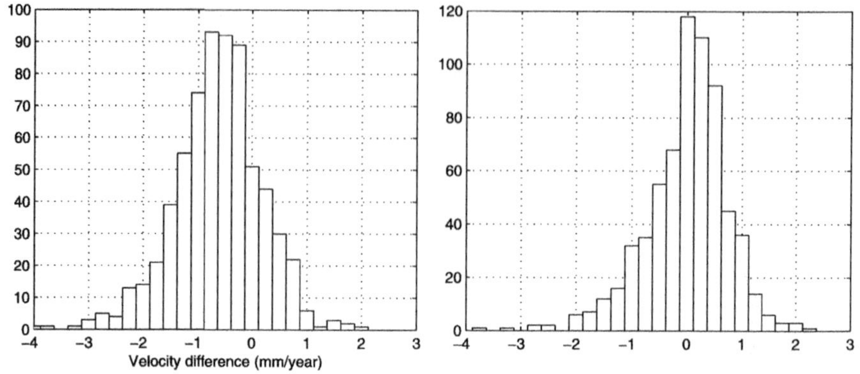

Fig. 7.13 Histograms of the differences between PSI and leveling displacement rates: PSI—leveling (mm/year). Both before (left) and after correction for a bias and spatial trend (right). The histograms are slightly skew-symmetric. This may be caused by an additional component in the PS velocity estimates in the areas with a deviating behavior in Fig. 7.12. This can for example be caused by superposed deformation regimes (e.g. shallow compaction), or horizontal PS movements

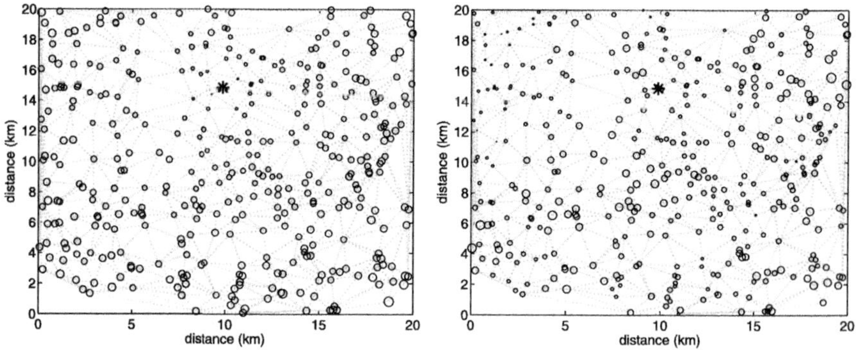

Fig. 7.14 Simulated network of uncorrelated height measurements per arc. Left: standard deviations of the estimated heights. Right: actual height estimates. The reference point is depicted as a black star. A larger marker size indicates a higher standard deviation (left) or height estimate (right). It can be seen that a spatial trend in the height estimates (right) can be introduced by correlation due to the network design

Before looking into the physical causes of the differences, the stochastic nature of both PSI and leveling is considered. In the comparison of the overlapping PSI tracks in Sect. 6.4, a spatial trend of several mm/year over 100 km was recognized. The existence of spatial trends in the results is not restricted to PSI: the leveling heights can exhibit spatial trends as well. This is due to the error propagation in the leveling network: nearby benchmarks are dependent on the measured height differences of almost similar network paths. As an example, Fig. 7.14 shows the correlation of the height estimates in a simulated leveling network. All height differences are 0; the measurement noise has been set to 1 mm/$\sqrt{\text{km}}$. It can be seen that, although the

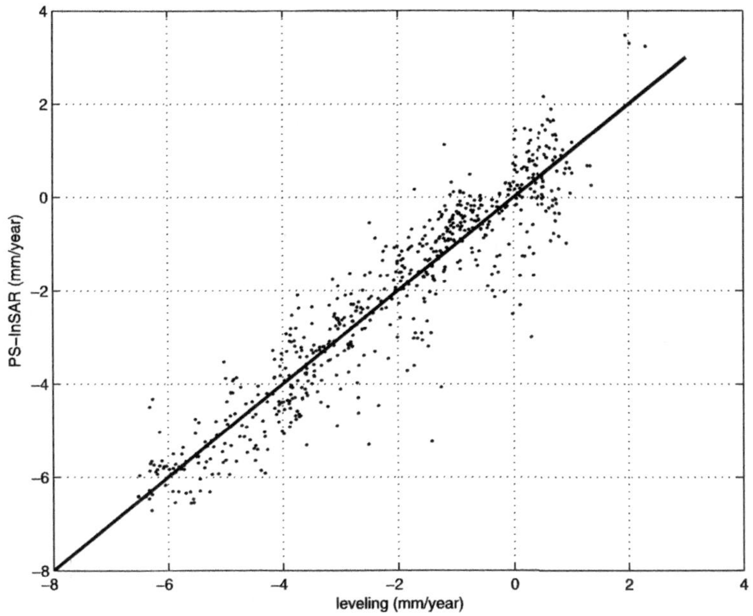

Fig. 7.15 Scatterplot of leveling versus PSI displacement rates after correction for a bias and a spatial trend. The correlation coefficient is 0.94

height difference observations are uncorrelated, the estimated heights are correlated due to the network design. The heights in the bottom right corner are relatively larger than the heights in the top left corner. This implies that the spatial trend from Fig. 7.12 can be caused by both PSI and leveling.

Hence, it is valid to correct for a bias and a spatial trend to determine the correlation coefficient between leveling and PSI. Figure 7.15 depicts the correlation between the leveling and the PSI displacement rates. The correlation coefficient is 0.94. Recall that in the corner reflector experiment a correlation coefficient of 0.94 was obtained between leveling and InSAR double-difference displacements. The correlation coefficient between leveling and PSI displacement rates is therefore approximately equal to the correlation that can be achieved in a controlled experiment. Moreover, the correlation coefficient of displacement estimates from repeated leveling campaigns is also lower than 1 ($\sim$0.94–0.97), due to the measurement precision, see Sect. 7.2.3. Hence, the correlation between PSI and leveling displacement estimates reaches its maximum.

7.2.2.2 Potential Causes for the Differences between PSI and Leveling

Although the correlation is high, it can be seen from Fig. 7.12 that areas can be indicated that exhibit systematically higher PS displacement rates compared to the leveling displacement rates. Although the differences are minimal, potential causes are summarized in this section.

A possible hypothesis is the existence of an additional compaction component in areas where the majority of the buildings and structures have a shallow foundation (Schroot et al., 2003). In such a situation, the number of PS that refer to well founded objects are a minor subset of the total amount of PS. Hence they are considered as outliers, and are subsequently removed. As a result, the average displacement rate of the remaining PS contains an additional compaction component. In practice, the validation of this hypothesis requires identification of the PS origin and its foundation with respect to the subsurface layers. Because such a procedure is labor intensive, it should first be determined if the differences between leveling and PSI displacements are significant. This will be addressed in Sect. 7.2.3.

Another explanation for the differences between the PSI and leveling displacement rates can be found in the spatial decomposition. Currently, the PS displacement rates are converted from satellite line of sight to the vertical. In this conversion, horizontal components are neglected. The errors that are introduced by disregarding the horizontal components have been evaluated for a simplified representation of the Groningen gas field. Figure 7.16 shows the expected horizontal and vertical displacement rates for a 170 meters thick disc-shaped reservoir at 3 kilometers depth with a radius of 15 km (Geertsma, 1973b). The pressure drop in 10 years is ~36 bar; the compaction coefficient $0.72 \cdot 10^{-5}$ bar^{-1}. It can be seen that in the presence of horizontal movements, the vertical and the line of sight displacements differ in the order of −0.5 up to 1 mm/year. If line of sight displacement rates are converted under the assumption that PS movements are only vertical, deviations of ~1 mm/year are found with respect to the actual vertical movements. The largest errors occur where the slope of the subsidence bowl is the steepest. Here, the horizontal movements due to gas extraction reach their maximum.

To conclude, the differences between PSI and leveling displacement rates can be caused by errors in the PSI estimation, such as unwrapping errors. However, since

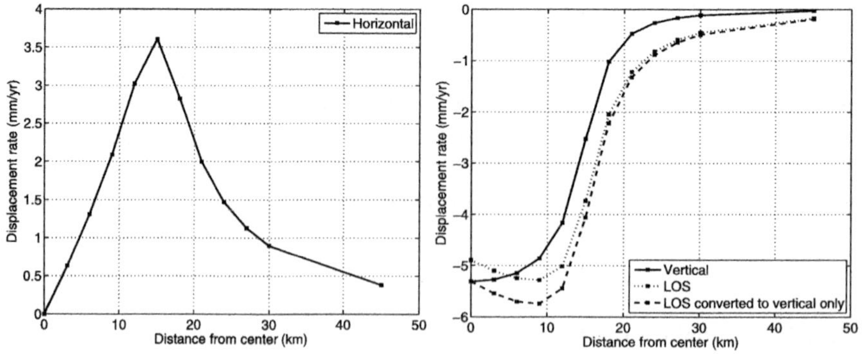

Fig. 7.16 Maximum expected horizontal displacement rates in the Groningen subsidence area (left). If line of sight displacement rates are converted to the vertical without taking horizontal movements into account, errors are ~1 mm/year (right)

the PSI results stem from the datum connection procedure of six independent tracks, this cause is less likely.

7.2.3 *PSI and Leveling Displacements*

This section compares leveling and PSI *displacements* in a fixed time period. The first reason for this comparison is the fact that displacement rates are not necessarily constant in time. If the gas production rates change significantly, the displacement rates will follow. The second reason is that subsidence due to gas extraction is reported as the total amount of subsidence since the start of the production. Hence, the total displacements between subsequent leveling epochs are compared. After explanation of the comparison set up, the results are discussed.

7.2.3.1 Setup of the Comparison

The start and end of the periods over which the displacements are compared, coincide with the leveling epochs. The main leveling epochs that cover the period of PSI measurements are 1993, 1998 and 2003. Hence, the leveling and PSI displacements have been compared in the intervals 1993–1998 and 1993–2003 (ERS).

The deformation estimates of both leveling and PSI are spatially relative. Due to a different reference point, there will be a constant offset between the PSI and the leveling results. Moreover, as discussed in Sect. 7.2.2.1, an additional spatial trend can be present in both the leveling and the PSI results. It has been chosen to correct the PSI displacement estimates for a bias and a spatial trend. The spatial trend in the PSI estimates is estimated by utilizing PSI estimates that cover the non-deforming areas, independent from the leveling displacement estimates.

To benefit from the high temporal sampling of PSI, the displacements between two dates are estimated assuming linear displacements in a fixed time window (according to the leveling campaign intervals).

The procedure to prepare PSI estimates for the comparison with leveling starts with the conversion of PSI displacements to the vertical. Subsequently, the displacement *rates* in a fixed time window are estimated. From these displacement rates, the displacements in the time window are estimated. An additional step corrects the PSI estimates for one offset and a spatial trend. It has to be noted that ERS-2 lost its three gyro mode in the beginning of 2000, and the number of useful acquisitions has significantly dropped afterwards. Therefore, the estimated PSI displacement rates for 1993–2003 mainly depend on the InSAR observations between 1993 and 1999.

Subsequently, leveling and PSI displacements are compared at evaluation locations along a certain trajectory (profile). Profiles have been defined along different orientations covering the Groningen subsidence bowl. To benefit from the spatial PS density, weighted averages of the PS displacements within a radius of one kilometer with respect to the evaluation locations have been computed.

Due the amount of PSI estimates, the comparison is currently based on the reduced datasets that consist of a selection of the PS with the highest precision within grid cells of 100×100 m. Furthermore, a data snooping procedure has been applied to remove gross outliers.

7.2.3.2 Results

The results of the PSI and leveling displacement comparison at evaluation locations along the profiles are depicted in Appendix 2. Two estimation methodologies for the leveling displacements have been applied. The first methodology computes the leveling displacements from the height estimates that are obtained from the free network adjustments per epoch (Teunissen, 2000a). The second methodology uses the Subsidence Modeling concept (Odijk and Kenselaar, 2003) that estimates the spatially correlated deformation signal. Moreover, SuMo removes outliers, identification errors and benchmarks that exhibit autonomous movements, due to the integrated spatio-temporal approach. Identification errors and autonomous movements cannot be detected in the free network adjustment of a single epoch, and hence explains the spikes in the leveling displacement profiles, see Figs. A.6, A.7, A.8, and A.9. This stresses that also leveling displacement estimates are dependent on the processing methodology.

The match between the leveling and the PSI displacement estimates can be quantified using the precision measures of both techniques, that have been defined in the following way:

- free network adjustments (leveling): propagation law of variances and covariances, considering the epochs mutually uncorrelated,
- SuMo (leveling): 2 mm/$\sqrt{\text{yr}}$ ('model imperfections'),
- PSI: standard deviation of the weighted average of the displacement estimates per evaluation location, taking the correlation of double-difference observations into account.

The chosen precision measure for the SuMo displacement estimates represents the uncertainty in time of the spatially correlated deformation signal.

The leveling and PSI profiles in the periods 1993–1998 and 1993–2003 match within the PSI and leveling error bounds, see also the distribution of the teststatistics in Figs. A.4 and A.12. Moreover, it is important to note that the transition to another sensor appears not to affect the continuity of subsidence monitoring, see Fig. 7.17. This figure shows the PSI and leveling displacement estimates (extrapolated) in the period 2003–2007 at equally spaced locations along a profile. Besides the continuity of subsidence monitoring, it shows the strength of PSI due to the high spatial and temporal point density: PSI can provide additional displacement estimates in areas where leveling benchmarks are not present.

Since the teststatistics per evaluation location are not uncorrelated, the overall model test is performed additionally, using (7.28). Since leveling epochs can be considered uncorrelated, $Q_{d_{\text{lev}}}$ can be computed by the addition of the variance-covariance matrices of the height estimates from the leveling epochs that have been

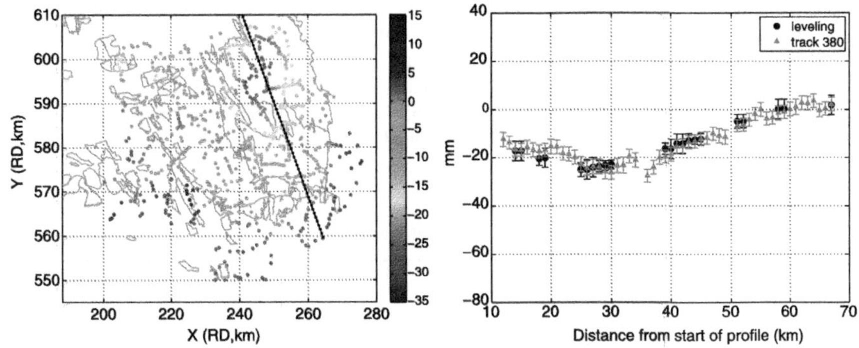

Fig. 7.17 Leveling and PSI displacements (mm) in the period 2003–2007 based on equally spaced locations along the profile (left). Right: Envisat track 380, and SuMo leveling displacements; PSI search radius 1 km. It can be deduced that the high spatial density of PSI can be utilized in areas where leveling benchmarks are not present. The leveling displacements are based on an extrapolation using the leveling displacement estimates in the period 1993–2003

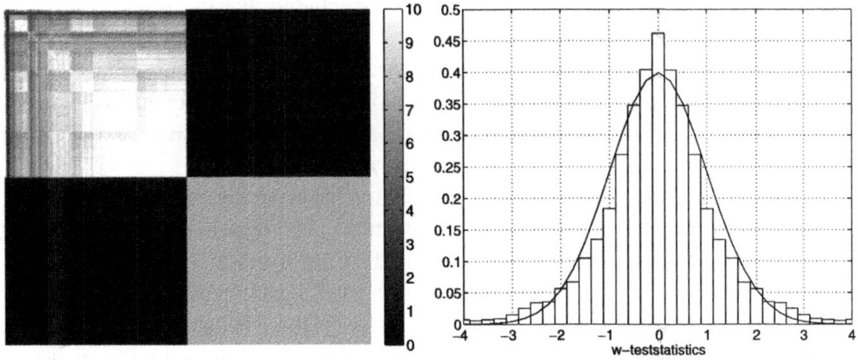

Fig. 7.18 Left: variance-covariance matrix of the mathematical model for the overall model test of (7.28). The top-left part of the matrix depicts the variance-covariance matrix of the leveling displacement estimates. The bottom-right part depicts the substitute matrix for PSI. The substitute matrix for PSI is a full matrix due to the double-difference combinations and is a simplified representation. Right: the $\underline{w}$-teststatistics and their theoretical distribution. After outlier removal, the overall model teststatistic is 1.05 and hence the null hypothesis is accepted, i.e. PSI and leveling displacements are in agreement

used to compute the displacements. For PSI, $Q_{d_{\text{PSI}}}$ is currently constructed as a substitute matrix, since the PSI deformation estimates are performed per arc. The substitute variance-covariance matrix takes into account the correlation between the double-difference displacements and accounts for the different precision (weights) of the PSI displacement estimates. The variance-covariance matrix that has been used for the overall model test is depicted in Fig. 7.18. The PSI variance-covariance matrix can be further refined by incorporating stochastically modeled residual atmospheric signal and unmodeled deformation. For the period 1993–1998, the over-

Table 7.1 Correlation coefficients in the comparison of PSI and leveling displacement estimates in the periods 1993–1998 and 1993–2003, at all leveling benchmark locations and at the subset of geostatistically and physically stable benchmarks only (Schoustra, 2006). Correlation coefficients have been computed for the leveling displacements estimated from both the free network adjustments and the SuMo analysis. The PSI displacements have been corrected for a spatial trend based on stable areas in the large coverage of PSI, independent of the leveling displacement estimates. The PSI and leveling displacements have been corrected for one constant offset, due to a different reference point

	all benchmarks	benchmark selection
SuMo analysis 1993–1998	0.90	0.93
SuMo analysis 1993–2003	0.91	0.95
Free network adjustments 1993–1998	0.74	0.87
Free network adjustments 1993–2003	0.81	0.94

all model teststatistic is 4.29 and 1.05 before and after outlier removal (4.5% of the observations) respectively. The critical value k_α is 1.07 and 1.13 for $\alpha = 0.05$ and $\alpha = 0.001$ respectively. Hence, it can be concluded that the PSI and leveling displacement estimates are in agreement, after outlier removal. The outliers can be present both in the PSI and the leveling displacement estimates. Figure 7.18 depicts the histogram of the $\underline{w}$-teststatistics (datasnooping) and their theoretical distribution. As expected—since the null hypothesis is accepted—there is a good agreement between the histogram and the theoretical distribution.

Figure A.3 and Fig. A.11 show the correlation between the PSI and the leveling displacement estimates at the benchmark locations. The obtained correlation coefficients in the periods 1993–1998 and 1993–2003 are listed in Table 7.1. The correlation coefficients are listed for an evaluation at all benchmark locations, and for a sub-selection of geostatistically and physically 'stable' benchmarks (Schoustra, 2006). The maximum correlation coefficient between the leveling and PSI displacements is 0.93–0.95.

For comparison, the obtainable correlation coefficients for repeated leveling campaigns have been determined by means of simulations. The height difference measurements have been simulated as a superposition of a deterministic part based on the subsidence prognoses (prediction based on geomechanical modeling), and a random part based on the measurement precision ($\sim 1\ \mathrm{mm}/\sqrt{\mathrm{km}}$), using the network designs of the existing leveling campaigns. Figure 7.19 shows the histograms of the correlation coefficients obtained for simulations of the displacements in the periods 1993–1998 and 1993–2003, in the Groningen subsidence area. The correlation varies from $\sim$0.94 for the period 1993–1998 to $\sim$0.97 for the period 1993–2003. The higher correlation coefficient for the period 1993–2003 is caused by the larger magnitude of the deformation signal. Since the leveling measurement precision remains unchanged, the linear relationship between the estimated displacements in the various simulations gets stronger when the range of displacements increases.

The correlation coefficient of 0.94–0.97 for deformation estimates from repeated leveling campaigns only represents the variability due to the measurement precision.

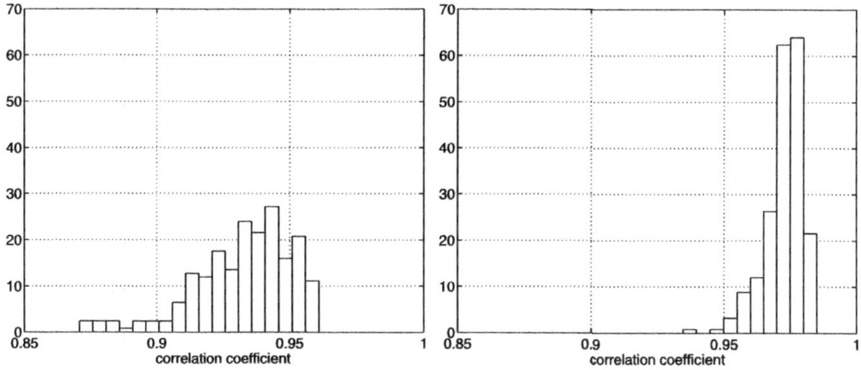

Fig. 7.19 Correlation coefficients between deformation estimates obtained from simulated leveling campaigns in the period 1993–1998 (left) and 1993–2003 (right). The leveling height observations have been simulated as a superposition of a deterministic signal (the subsidence prognosis) and the measurement precision. The correlation coefficients only represent the variability due to the leveling measurement precision; benchmark instabilities are not considered. Since the measurement precision stays unchanged and the magnitude of the subsidence signal increases in time, the correlation coefficient is larger for the period 1993–2003 (~0.97) than for the period 1993–1998 (~0.94)

In practice, it is expected that the correlation coefficient will be slightly lower due to benchmark instabilities. It can be concluded that the correlation between PSI and leveling displacement estimates is similar to the correlation between displacement estimates from repeated leveling campaigns. This means that PSI has reached the maturity to be operationally used for subsidence monitoring in the northern part of the Netherlands.

7.3 The Integration of Geodetic Measurement Techniques

In Sect. 7.1, the application of leveling and PSI has been considered from a theoretical perspective taking the spatio-temporal observation frequency and the measurement precision into account. After it has been demonstrated that the spatio-temporal observation frequency can overcome the disadvantage of a relatively lower measurement precision, displacement rates and displacements in a fixed time period have been compared in Sect. 7.2. The correlation coefficient of leveling and PSI displacement rates (0.94) is similar to the correlation coefficient of leveling and InSAR displacements in a controlled corner reflector experiment (0.94). Moreover, the misclosure teststatistics between leveling and PSI match their theoretical distribution. This implies that leveling and PSI measurements can be integrated for subsidence monitoring. Hence, this section proposes a mathematical framework for the integration of multiple (geodetic) techniques that observe the same deformation signal of interest. It introduces theoretical guidelines for deformation monitoring using multiple techniques. Although similar concepts can be found in Chap. 3 and 4, the integrated approach has not yet been applied in practice for monitoring subsidence due to gas extraction.

7.3.1 Mathematical Model

To develop a unified strategy for the integration of observations from multiple techniques, the following has to be taken into consideration:

1. type of observations,
2. measurement precision,
3. parameterization of deformation signal, and
4. idealization precision for deformation monitoring of the signal of interest.

Prior to addressing these four issues, the mathematical framework for deformation monitoring using M measurement techniques is introduced:

$$\begin{bmatrix} \underline{y}_1 \\ \vdots \\ \underline{y}_M \end{bmatrix} = Ax + \begin{bmatrix} \sum_{d=1}^{D_1} \underline{s}_d(x,y,t) \\ \vdots \\ \sum_{d=1}^{D_M} \underline{s}_d(x,y,t) \end{bmatrix} + \begin{bmatrix} \underline{n}_1 \\ \vdots \\ \underline{n}_M \end{bmatrix}, \tag{7.30}$$

where:

$\underline{y}$	measurement input vector
A	design matrix that defines the relation between the measurement input and the unknown deformation parameters,
x	unknown deformation parameters,
$\underline{s}_d(x,y,t)$	signal that describes the discrepancy between modeled and actual deformation for deformation regime d with a certain spatial (x,y) and temporal (t) behavior,
$\underline{n}$	measurement error.

This system of equations can be reformulated as a combined functional and stochastic model:

$$E\{\underline{y}\} = Ax; \qquad Q_y = \sum_{d=1}^{D} Q_{ss_d(x,y,t)} + \begin{bmatrix} Q_{nn_1} & 0 & 0 \\ 0 & \ddots & 0 \\ 0 & 0 & Q_{nn_M} \end{bmatrix}, \tag{7.31}$$

where D is the total number of deformation regimes observed by all measurement techniques together. The measurement precision that is represented by Q_{nn} is assumed to be uncorrelated between the different measurement techniques.

The measurement input can consist of observations or deformation estimates. The latter implies an estimation in phases. This occurs for example when PSI and leveling measurements are integrated for the estimation of subsidence due to gas extraction. First, deformation is estimated from the interferometric phase difference observations. These deformation estimates contain displacements due to all superposed deformation regimes. Secondly, these deformation estimates are integrated with leveling observations to estimate the actual deformation due to gas extraction only. The composition of the measurement input does not affect the outcome of the estimation of the deformation signal of interest, provided that the stochastic information is preserved. This means that the variance-covariance matrix of the deformation estimates in the first step is required as input for the estimation of the deformation signal of interest.

Regardless whether the measurement input consists of observations or deformation estimates, their nature has to be considered for each measurement technique:

- absolute or relative displacements,
- orthometric or ellipsoidal displacements.

Absolute displacements can be measured by gravity (displacements orthogonal to the potential field) or GPS (coordinates in the International Terrestrial Reference System (ITRS)). The majority of the measurement techniques, such as leveling and PSI, provide relative observations. These relative observations can be further subdivided based on their spatio-temporal contents. An interferometric phase observation is a temporal difference, whereas a leveling measurement is a spatial height difference between two benchmarks, see Fig. 7.1. The first interpretable PSI measurement is the double-difference observation. As explained in Sect. 7.1.1, leveling and PSI observations can be converted to double-differences for comparison. The number of independent double-differences that can be formed depends for leveling on the network design, whereas for PSI always $(P - 1)$ double-differences can be formed from P PS. Furthermore, orthometric and geometric (ellipsoidal) displacements are distinguished, as described in Sect. 7.2.2.

The measurement input precision is described by the stochastic model of the observations or the deformation estimates that are utilized for the estimation of the deformation signal of interest. The stochastic model comprises all stochastically modelled error sources of a measurement technique. For leveling, it is restricted to measurement noise ($\sim$1 mm/$\sqrt{\text{km}}$). For space borne techniques such as PSI it includes stochastically modeled errors, such as atmospheric disturbances.

Parameterization of the deformation signal depends on the signal of interest. The options for the functional modeling of subsidence due to gas extraction have been discussed in Sect. 2.3.3.1. It varies from modeling subsidence due to gas extraction by a point source to a prognosis grid based on a geomechanical model of the subsurface. Here, the multi-disciplinary approach of deformation modeling comes in. The precision of the prognosis can be optimized by modeling the geomechanical properties of the subsurface.

To conclude, the idealization precision for deformation monitoring describes how well the deformation signal of interest can be monitored by a certain measurement technique. It is a combination of the physical identification level of the measurement points and the knowledge on the behavior of the subsidence signal of interest in the potential presence of other deformation regimes.

7.3.2 The Integration of Leveling and PSI

In this section, the mathematical framework that has been introduced in Sect. 7.3.1 is applied more specifically to the integration of leveling and PSI deformation estimates. Because of the different physical properties of the measurement points, benchmark versus reflection, it is not possible to compare the observations directly. The deformation estimates of each technique are integrated in the parameter space: the joint estimation of the deformation signal of interest. The measurement inputs in this joint estimation are the leveling and PSI deformation estimates.

The PSI deformation estimates are estimated from the PSI system of equations. Recall the redundant parameter estimation after unwrapping of the phase observations from (6.4):

$$
\begin{aligned}
E\{\underline{y}\} = E\left\{\begin{bmatrix} \underline{\varphi}_{ij}^{k=1} \\ \vdots \\ \underline{\varphi}_{ij}^{k=K} \end{bmatrix}\right\} &= \begin{bmatrix} -\frac{4\pi}{\lambda}T^k & -\frac{4\pi}{\lambda}\frac{B_i^{\perp}}{R_i^m \sin\theta_i^m} \end{bmatrix}\begin{bmatrix} v \\ H \end{bmatrix}; \\
D\left\{\begin{bmatrix} \underline{\varphi}_{ij}^k \end{bmatrix}\right\} &= Q_{nn} + Q_{\text{atmo}} + Q_{\text{defo}}.
\end{aligned}
\tag{7.32}
$$

The PSI deformation estimates that are the measurement input in the joint estimation of the signal of interest can be determined using the Best Linear Unbiased Prediction (BLUP) theory, see e.g., Teunissen et al. (2005). Here, a realization of the phase contributions due to deformation is predicted, including a corresponding variance-covariance matrix that only depends on the uncertainty in the deformation modeling. Hence, the PSI system of equations is extended in the following way:

$$
E\left\{\begin{bmatrix} \underline{y} \\ \underline{z} \end{bmatrix}\right\} = \begin{bmatrix} A \\ A_z \end{bmatrix} x,
\tag{7.33}
$$

more specifically:

$$
E\left\{\begin{bmatrix} \underline{\varphi}_{ij}^k \\ \underline{z} \end{bmatrix}\right\} = \begin{bmatrix} -\frac{4\pi}{\lambda}T^k & -\frac{4\pi}{\lambda}\frac{B_i^{\perp}}{R_i^m \sin\theta_i^m} \\ -\frac{4\pi}{\lambda}T^k & 0 \end{bmatrix}\begin{bmatrix} v \\ H \end{bmatrix},
\tag{7.34}
$$

with the stochastic model:

$$
D\left\{\begin{bmatrix} \underline{\varphi}_{ij}^k \\ \underline{z} \end{bmatrix}\right\} = \begin{bmatrix} Q_{yy} & Q_{yz} \\ Q_{zy} & Q_{zz} \end{bmatrix} = \begin{bmatrix} (Q_{nn} + Q_{\text{atmo}} + Q_{\text{defo}}) & Q_{\text{defo}} \\ Q_{\text{defo}} & Q_{\text{defo}} \end{bmatrix}.
\tag{7.35}
$$

The vector $\underline{z}$ contains the predicted phase contribution due to deformation:

$$
\hat{\underline{z}} = -\frac{4\pi}{\lambda}T^k\hat{v} + \hat{\underline{s}}_{\text{defo}},
\tag{7.36}
$$

where $\hat{\underline{s}}_{\text{defo}}$ is the phase contribution due to unmodeled deformation signal. Applying the propagation law to the predicted deformation vector results in its error variance-covariance matrix:

$$
P_{\hat{z}\hat{z}} = Q_{zz} - Q_{zy}Q_{yy}^{-1}Q_{yz} + (A_z - Q_{zy}Q_{yy}^{-1}A)Q_{\hat{x}\hat{x}}(A_z - Q_{zy}Q_{yy}^{-1}A)^T,
\tag{7.37}
$$

where:

$$
\begin{aligned}
A &= [-\tfrac{4\pi}{\lambda} T^k \quad -\tfrac{4\pi}{\lambda} \tfrac{B_i^\perp}{R_i^m \sin\theta_i^m}], \\
A_z &= [-\tfrac{4\pi}{\lambda} T^k \quad 0], \\
Q_y &= Q_{nn} + Q_{\text{atmo}} + Q_{\text{defo}}, \\
Q_{zz} &= Q_{\text{defo}}, \\
Q_{zy} &= Q_{\text{defo}}, \\
Q_{yz} &= Q_{\text{defo}}.
\end{aligned}
$$

The advantage of using the predicted deformation vector instead of the PSI deformation estimates, is that the uncertainty described in its variance-covariance matrix is only due to the stochastic uncertainty in the deformation signal. The variance-covariance matrix of the PSI deformation estimates also contains contributions due to measurement noise and atmospheric disturbances. However, there are also disadvantages of using the predicted deformation vector. The main complicating factor is, that the covariance functions of the measurement noise, (residual) atmospheric signal, and unmodeled deformation have to be realistic. Otherwise, there is a risk of incorrectly addressing contributions due to unmodeled deformation to other error sources. In such a situation, the use of the PSI deformation estimates together with the variance-covariance matrix containing the contributions of all error sources should be preferred.

This mathematical framework for the joint estimation of a deformation signal of interest using multiple measurement techniques has not yet been implemented. For the estimation of subsidence due to gas extraction, it would certainly require a multi-disciplinary approach. The geomechanically modelled subsidence prognosis including its uncertainties has to be included in an iterative procedure to obtain agreement with the geodetic measurements at surface level.

7.4 Conclusions

This chapter has investigated the operational use of PSI for subsidence monitoring. Therefore, the PSI results have been compared to the subsidence estimates obtained from leveling campaigns in the Groningen area. This comparison has been performed from an integrated perspective, taking the uncertainty of both techniques into account. It has been shown by means of simulations, that the spatio-temporal point density of PSI can lead to an equal or even higher precision of displacement rate estimates compared to the leveling technique, although the precision of PSI double-difference observations is lower, see Fig. 7.3.

The temporal sampling of PSI has been evaluated using the DOP precision measure for the variance-covariance matrix, which is independent of the spatio-temporal reference. Both for PSI and leveling, a monitoring period of 10 years has been chosen. Considering PS double-difference displacements with standard deviation of 3 mm, and leveling height difference observations with a precision of 1 mm, $\sim$25 SAR acquisitions are required to obtain the same precision for the displacement rates.

In the evaluation of the spatial sampling, both the point density and the smoothness of the deformation signal of interest have been investigated. In the evaluation of the spatial point density, it is important that the correlation among both leveling and InSAR double-difference displacements are taken into account. Otherwise, the precision of the deformation estimates due to a higher spatial sampling are overestimated. Furthermore, the spatial sampling to reconstruct the subsidence signal in the Groningen area has been investigated, using a cross-validation procedure of the actual subsidence prognosis and the interpolated deformation signal from samples with varying point density. It could be concluded that a point density of $\sim$1 per km^2 is required to reconstruct the subsidence signal in the Groningen area.

After showing that PSI can compete with leveling based on its spatio-temporal density, the comparison of the actual PSI and leveling displacement estimates has been performed in the period 1993–2003. The correlation between the displacement rates of both techniques is 0.94, which is comparable to the correlation of the leveling and InSAR displacements of the controlled corner reflector experiment (0.94). This correlation coefficient reaches its maximum, since the correlation coefficient of displacement estimates from repeated leveling campaigns is also lower than 1 ($\sim$0.94–0.97), due to the measurement precision.

The displacements have been compared in fixed periods (1993–1998 and 1993–2003) at evaluation locations along profiles that cover the Groningen subsidence bowl. It has been shown that the misclosure teststatistics agree with the theoretical distribution. The maximum correlation coefficients between the PSI and leveling displacements are 0.94–0.95. A few areas can be pointed out in which the PSI subsidence rates are slightly larger than the leveling subsidence rates. Although the causes for these minor deviations have to be further investigated, it can be stated that PSI is ready for operational use for monitoring subsidence due to hydrocarbon production, considering the correlation between leveling and PSI and the multi-track reliability analysis from Chap. 6. To conclude, a mathematical framework has been introduced for a rigorous integration of multiple measurement techniques.

Chapter 8
Discussion and Future Subsidence Monitoring

In this chapter, the obtained results in the Groningen area from the ERS-1, ERS-2 and Envisat missions are summarized and discussed. Both the precision and reliability of PSI as a measurement technique, and the idealization precision for monitoring subsidence due to gas extraction, are addressed. Additionally, examples of the use of PSI to increase insight in reservoir behavior are shown. To conclude, a future scenario for subsidence monitoring is proposed.

8.1 Precision and Reliability

PSI observations are double-difference phase observations. They refer to a spatial and temporal reference: one PS and one acquisition time. The precision of the PS deformation estimates is independent of the reference PS. Using the variances of the PS deformation estimates as an absolute precision measure, suggests that PS at a larger distance from the reference PS have a lower precision. However, the relative precision, described by the full variance-covariance matrix, is invariant. To parameterize the precision represented by the variance-covariance matrix independent of the reference PS, the Dilution of Precision (DOP) measure can be applied, see Sect. 4.3.4.

A validation of the stochastic model for InSAR has been performed in a controlled corner reflector experiment, see Sect. 4.4. Utilizing the independent leveling technique, the precision of PSI double-difference displacements has been estimated by means of variance component estimation. The estimated precision (1-sigma) for ERS-2 and Envisat double-difference displacements is 3.0 and 1.6 mm respectively. The lower precision of the ERS-2 double-differences is likely to be caused by the large Doppler centroid frequency deviations in the time series that cover the period 2003–2007. The correlation between the leveling and Envisat double-difference displacements is 94%, which demonstrates the potential of InSAR as a deformation monitoring technique.

After validation of the measurement precision of InSAR in the controlled corner reflector experiment, the precision of deformation estimates obtained from natural PS has been investigated. In the absence of unwrapping errors and other systematic errors, the precision of the PSI deformation and height estimates is determined by the measurement precision, the physical PS properties, and the network design, i.e. temporal acquisition density and viewing geometry. The precision of the ERS and Envisat PS velocities show a dependency on the distance between the PS, see Sect. 6.3.1. The smallest stack, that consists of 24 interferograms over a period of 8 years, has a PS velocity precision of $\sim$0.1 mm/year per $\sqrt{\text{km}}$. The largest stack, that consists of 74 interferograms that cover a period of 14 years, exhibits a PS velocity precision of $\sim$0.04 mm/year per $\sqrt{\text{km}}$.

V.B.H. (Gini) Ketelaar, *Satellite Radar Interferometry*,
Remote Sensing and Digital Image Processing 14,

Unmodeled additional error sources and unwrapping errors can lower these precision values. First of all, neglectance of the azimuth sub-pixel position can lead to an additional error of ~0.5 mm/year in the PS velocity estimates, see Sect. 4.2.1. Furthermore, orbit errors have to be considered, since the Groningen subsidence area has a large extent. Random orbit errors of 5 and 8 centimeters in radial and across-track direction can lead to velocity errors up to ~1 mm/year between near and far range, see Sect. 4.2.3.

Since the main deformation signal exhibits low subsidence rates (<7 mm/year) and the temporal sampling of the near-linear displacements is relatively high (up to 10 acquisitions per year per track), the success rates for phase ambiguity resolution approach 1 for high PS densities (>100 PS/km^2) and a high phase precision (1/20 cycle, 1.5 mm for displacements), see Sect. 4.2.4. However, since the PS density in rural areas is limited to 0–10 PS/km^2, atmospheric disturbances can deteriorate the phase unwrapping success rates. From simulations, it could be deduced that when using a sparse network of arcs with high measurement precision, high success rates can be obtained (>0.9), but that the success rates drop with lower measurement precision (1/10 cycle) and PS density (5 PS/km^2), see Sect. 4.2.4. Ambiguity resolution success rates of 1 can therefore not be guaranteed in the rural Groningen area, even not after removal of misclosures in the test procedure for spatial unwrapping, see Sect. 6.1.3. Denser PS networks may lead to a better discrimination in the determination whether a PS candidate should be accepted or not, but does not solve the problem that the PS parameter estimates cannot be tested due to the lack of redundancy in the system of equations, see Sect. 3.4.1. One unwrapping error in the time series for the Groningen acquisition geometry and sampling can lead to errors in the order of 1 mm/year. The success rate of unwrapping decreases in case of deformation model imperfections. An example is shown in Fig. 8.3: the estimation of a linear displacement rate in combination with the data gap around 1994 results in a high likelihood for the displacements in 1992–1993 to have an offset of −28 mm (half a wavelength).

To introduce redundancy for a reliability assessment, multiple independent overlapping tracks have been utilized that monitor the same deformation signal. The Groningen subsidence bowl is visible in six overlapping tracks. A datum connection procedure has been developed to integrate these tracks providing a reliability assessment at the same time, see Sect. 5.2. After conversion to a common radar datum, multi-track PS within limited or even resolution cell distance have been unambiguously detected. The misclosures between PSI estimates (displacements, heights) of different tracks should theoretically consist of a constant offset due to a different reference PS. However, it has appeared that small spatial trends of several mm/year over a 100 km distance are present in the PSI estimates. These trends can be caused by unwrapping errors and orbital inaccuracies that propagate over a large spatial extent. After datum connection, the standard deviation of PS velocity estimates is less than 1 mm/year for 70% of the PS clusters, see Sect. 6.4.1. Besides due to the precision of PSI as a measurement technique, these differences can be due to different deformation regimes (PS are physically not the same) and potential horizontal deformation components.

PSI deformation estimates along the line of sight from tracks with different viewing geometries can be further used for a decomposition into horizontal and vertical movements, see Sect. 6.4.2. It has been shown that the sign and magnitude of the horizontal components correspond with the theoretically expected horizontal movements for subsidence due to gas extraction. Although these horizontal components cannot be unambiguously subdivided into geophysical signal and residual systematic effects, it points out that the existence of horizontal components have to be taken into account. If one would neglect the horizontal components and convert line of sight estimates directly into vertical components only, an additional error component up to $\sim$1 mm/year is introduced. This stresses the importance of subsidence monitoring by an overlapping ascending and descending track.

8.2 Separation of Deformation Regimes

Deformation of the earth's surface can be driven by different mechanisms. Deformation causes can be classified into deformation regimes: structural instabilities, shallow subsurface movements and deep subsurface movements, see Sect. 4.5.1. Structural instabilities are settling effects of newly built buildings, or autonomous movements of buildings with an instable foundation. Shallow subsurface movements are natural compaction and movements due to groundwater level variations. Deep subsurface movements comprise gas, oil or other mineral extraction. Due to soft soils in large parts of the Netherlands, shallow subsurface movements cannot be neglected. Ground level movements due to shallow compaction can be very irregular, depending on the mechanism (groundwater level variations, peat oxidation, natural compaction). From external studies that have analyzed leveling benchmark movements in periods and areas that are not affected by gas extraction, it could be concluded that the majority of the movements are within $\pm$1 mm/year, with maximum displacement rates up to 1 cm/year, see Sect. 6.5.1. Corner reflector movements in the Delft experiment are approximately 1–2 centimeters seasonally, see Sect. 6.5.1. Since subsidence rates due to gas extraction in Groningen are maximum $\sim$7 mm/year, deformation due to shallow causes can potentially contaminate the deformation signal of interest.

Compared to traditional geodetic techniques, using fixed benchmarks, PSI measurement points are less well defined. Different reflection types (single, double, multiple-bounce), combined with a limited precision of the geographic location, makes it impossible to identify the exact physical properties of a Persistent Scatterer. Furthermore, since the radar satellite is monitoring from space, it measures all surface movements, regardless of the deformation regime. This further complicates the attribution of a PS displacement to a deformation regime. However, the identification of deformation regimes is also not straightforward for a land surveying technique such as leveling. Only few underground benchmarks are mounted on the stable Pleistocene layer, while the majority is mounted in buildings that may be potentially affected by additional autonomous movements. The measurements of both PSI and leveling potentially represent multiple deformation regimes. The difference is that in case of leveling the measurement point selection is performed during the network design ('filtering a-priori'). For PSI, the point selection in order to estimate the subsidence signal of interest is done after the estimation of deformation ('filtering a-posteriori'). However, even for leveling a-posteriori filtering is required, to distinguish benchmarks that exhibit autonomous movements from benchmarks that refer to the stable Pleistocene layer.

For a successful PSI estimation procedure, the PS density has to be sufficient, even in the rural areas of Groningen. It appeared that Persistent Scatterers in rural areas coincide with buildings (farms, houses) and other man-made structures, providing at least 0–10 PS per km^2 in 80% of the area, see Sect. 6.2.1. These Persistent Scatterers refer either to direct reflections from (the top of) the buildings, or to double-bounce reflections with respect to the ground level. The latter may possibly contain an additional shallow deformation component. Direct reflections from well-founded buildings are the most suitable observations to estimate subsidence due to gas extraction, and additional tools can be applied to select them: PS heights, Envisat's Alternating Polarization data, and reflection behavior as a function of viewing geometry (Perissin, 2006).

If PS heights are used for target characterization, it is important that sidelobe observations are removed. First of all because sidelobe observations are not independent, and secondly because sidelobe height estimates are biased because they refer to the wrong range bin. Furthermore, the PS heights have to be determined with respect to ground level, see Sect. 6.5.3. Since laser altimetry and SRTM data do not provide ground level heights in urbanized areas, local PS height histograms have been created. Based on the assumption that the majority of the PS represent double-bounce reflections from ground level, the ground level height corresponds with the location of the histogram peak. The uncertainty in the determination of the ground level heights is deduced by a histogram fit on the (multi-modal) local PS height histograms. For a case study, a standard deviation of the ground level heights of $\sim$2.5 meters was estimated, see Sect. 6.5.3. Subsequently, PS have been selected with a height above 5 meters, which have a probability of 95% to be elevated targets. Envisat Alternating Polarization data has confirmed that these reflections have a high probability to be odd-bounce (most likely specular).

To determine whether the selection of elevated specular PS targets influences the deformation estimates, the histograms of the PS velocities before and after selection have been compared. Since direct reflections from well-founded buildings should not be affected by additional shallow deformation components, it is to be expected that the PS velocity histogram shifts to displacement rates of a lower magnitude. It has appeared however, that the histogram shift is not significant in two case study areas, see Sect. 6.5.3. The histogram is indeed shifted to the stable area (meaning less additional autonomous and shallow compaction components), but differences are below 0.5 mm/year. The selection of direct reflections based on viewing geometry has resulted in similar PS velocity histogram differences, again not significant. Since subsidence due to gas extraction is the common deformation regime, the PSI results represent deep subsurface displacements, provided that the majority of the buildings does not move with respect to the Pleistocene layer.

For the estimation of subsidence due to gas extraction, PS can be selected based on the spatial correlation length of the deformation signal of interest, see Sect. 4.5.3 and Sect. 6.5.4. Benefitting from the high spatial PS density and the fact that gas extraction is the common deformation regime, a datasnooping procedure has been applied that removes all PS that are not spatially consistent with neighboring PS. The correlation coefficient between the spatially consistent PSI displacement rates and leveling displacement rates is 0.94, which approximates the correlation coefficient of displacements in a controlled corner reflector experiment (0.94), see Sect. 4.4.5. Considering the fact that the correlation coefficient between displacements estimated from repeated leveling campaigns is also not equal to 1 ($\sim$0.94–0.97), this suggests that PSI is ready for operational use for monitoring subsidence due to hydrocarbon production in the northern part of the Netherlands.

8.3 PSI and Reservoir Behavior

Acknowledging that subsidence due to gas extraction is successfully estimated by PSI, the question raises if the higher spatial and temporal observation density, compared to leveling, can be utilized to extend the knowledge on reservoir behavior. Especially in areas where drilling (additional) wells is considered, but high uncertainties exist about the behavior and connection of reservoir blocks, PSI may provide additional information. Another reason for the demand of increased reservoir knowledge is the start of subsidence due to gas extraction from the onshore side of the Waddenzee since February 2007. Since the Waddenzee gas production comprises several environmental concerns, subsidence due to gas production has to be monitored (near) real-time (NAM, 2006). The more knowledge on reservoir properties, the better the resulting subsidence can be controlled.

8.3.1 Temporal Behavior of Subsidence Due to Gas Extraction

Considering ERS and Envisat satellites, the temporal observation density can increase to maximum four observations in 35 days, if the area of interest is monitored from four independent tracks. Subsidence above the main Groningen gas field is near-linear in the satellite monitoring period of 1992–2007. However, some smaller fields have a different production history (NLOG, 2008). For example, the production of the Anjum field (see Fig. A.1) has started in the middle of the ERS monitoring period in 1997 (NAM, 2003b). Another example is the Norg field (see Fig. A.1), from which gas has been produced in the period 1983–1995 (NAM, 2003a). Successively, from 1997, it was used for underground gas storage, resulting in an uplift of the earth's surface.

An important reservoir parameter is the so-called subsidence delay: the time difference between the start of gas production and the onset of subsidence. The time

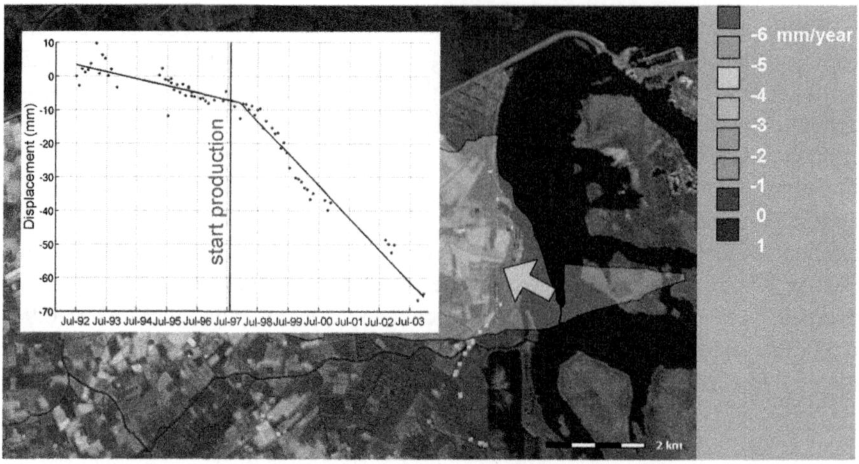

Fig. 8.1 Displacement time series of the Anjum gas field (in green). A change in displacement rate is observed several months to a year after the onset of gas production in August 1997. The average displacement rates are in mm/year. These displacement rates underestimate the subsidence rates after start of gas extraction, since they are estimated from the monitoring period 1993–2003. Source background map: Landsat 7. Overlay created in ArcGIS®

delay is closely linked to the shape of the subsidence curve until linear subsidence rates are reached at constant production rate (Hettema et al., 2002).

To preserve the PSI displacement series that deviate from the linear null hypothesis, the separation of unmodeled deformation signal and atmosphere (see Sect. 4.3.2) has been performed in a conservative way. The displacement time series may therefore be relatively noisy, but non-linearities in the time series are fully preserved. This section will demonstrate a method to detect the unmodeled deformation signal, based on the time series of displacement residuals.

Before investigating the displacement residuals, the uncertainties in the stochastic model of PSI need to be considered. The variance-covariance matrix of the observations consists of a superposition of measurement noise, atmospheric noise, and unmodeled deformation, see Sect. 3.4.3. PSI parameter estimation and variance component estimation have been performed per arc. Hence, atmospheric noise cannot be distinguished from measurement noise. The estimated variance factor therefore accounts for both measurement and atmospheric noise, acting as a scaling factor for different atmospheric disturbances (Hanssen, 2004). Figure 8.2 shows the estimated PS velocities and their estimated precision after VCE for a small study area that contains the Anjum gas field.

In Fig. 8.2, an area can be identified that shows a significantly lower PS velocity precision than the surroundings. Possible causes for the lower precision are a higher PS measurement noise, physical PS instability, or unmodeled deformation, i.e. the hypothesis of linear velocity is not valid. To trace model deviations, a residual analysis has been performed based on the least-squares residuals of the PS displacement estimates. The area is divided into blocks of 3×3 km, assuming a spatially smooth behavior of the signal of interest. From the residuals of all PS within a block, a

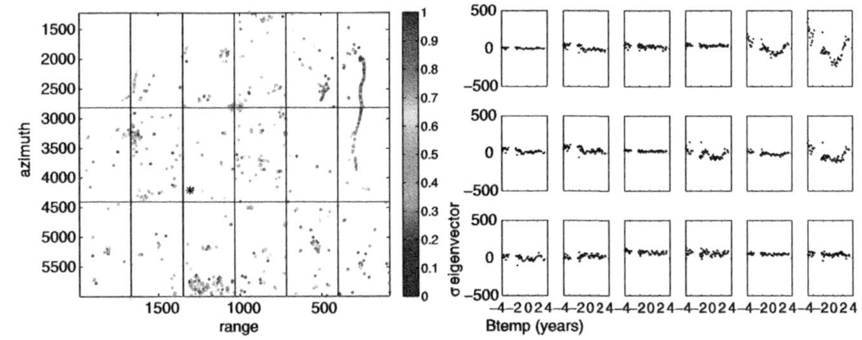

Fig. 8.2 PS velocity precision (mm/year) and division into blocks (left). The precision of the PS velocity estimates in the area in the top right corner is systematically lower compared to the surrounding PS. The reference PS is indicated by a black star. Right: the eigenvector per block that accounts for the largest variability in the residual PS displacements (with respect to a linear displacement rate). The blocks in the top right corner exhibit systematic deviations from the null hypothesis of a linear PS velocity. This is explained by the PS displacement time series in these areas, see Fig. 8.1

variance-covariance matrix is constructed:

$$C_{\hat{e}} = \underline{\hat{e}}\,\underline{\hat{e}}^T, \tag{8.1}$$

where $\underline{\hat{e}}$ is the matrix of least-squares residuals (number of interferograms × number of PS) and $C_{\hat{e}}$ represents the variance-covariance matrix of the residuals. This variance-covariance matrix is decomposed into eigenvectors and eigenvalues. For each block, the eigenvector corresponding to the largest eigenvalue, accounting for the largest variability within the block, is analyzed. From Fig. 8.2 it can be seen that the blocks with a lower precision show a systematic residual behavior that deviates from the linear velocity model. When looking into the PS time series, there is indeed a breakpoint in the displacement rate, see Fig. 8.1. Using the residual eigenvector analysis, areas can be traced where model assumptions have to be revised.

Alternative hypotheses can be defined to determine the most likely deformation pattern. One of the simplest alternative hypothesis consists of two linear displacement rates before and after start of gas extraction. More advanced functions for the (smooth) temporal development of subsidence due to mineral extraction are described by Kwinta et al. (1996).

Alternative hypotheses are evaluated using the overall model test (Teunissen, 2000b). An example of an alternative hypothesis is the extension of the deformation model with an additional unknown velocity and offset parameter u from time t:

$$H_0: \;\; E\{\underline{y}\} = Ax; \qquad H_a: \;\; E\{\underline{y}\} = Ax + C_y\nabla, \tag{8.2}$$

$$H_0: E\left\{\begin{bmatrix} \underline{y}_{ij}^{1} \\ \vdots \\ \underline{y}_{ij}^{t-1} \\ \underline{y}_{ij}^{t} \\ \vdots \\ \underline{y}_{ij}^{K} \end{bmatrix}\right\} = v_1 T^k;$$

$$H_a: E\left\{\begin{bmatrix} \underline{y}_{ij}^{1} \\ \vdots \\ \underline{y}_{ij}^{t-1} \\ \underline{y}_{ij}^{t} \\ \vdots \\ \underline{y}_{ij}^{K} \end{bmatrix}\right\} = \begin{bmatrix} T^t \\ \vdots \\ T^t \\ \vdots \\ 0 \end{bmatrix} v_1 + \begin{bmatrix} 0 & 0 \\ \vdots & \vdots \\ 0 & 0 \\ T^t & 1 \\ \vdots & \vdots \\ T^K & 1 \end{bmatrix} \begin{bmatrix} v_2 \\ u \end{bmatrix}. \tag{8.3}$$

Where v_1 and v_2 are the PS velocities before and after the start of subsidence due to gas production. The offset is required to avoid a discontinuity in the time series at the change of displacement rates. The corresponding teststatistics read:

$$H_0: \ \underline{T}_{q=m-n} = \hat{\underline{e}}_0^T Q_y^{-1} \hat{\underline{e}}_0; \qquad H_a: \ \underline{T}_{q=m-n} = \hat{\underline{e}}_a^T Q_y^{-1} \hat{\underline{e}}_a, \tag{8.4}$$

where $\hat{\underline{e}}$ are the least-squares residuals and Q_y is the variance-covariance matrix of the observations.

An example of a more complex deformation pattern is the displacement sequence of the Norg field, see Fig. A.1 and Fig. 8.3. The ground level is first subsiding due to gas production during the period 1983–1995 (NAM, 2003a). Subsequently, the Norg reservoir has been used for underground storage from 1997, resulting in an uplift of four centimeters in approximately four years.

Although the results clearly depict the displacements above the Norg field, it has to be noted that displacements that deviate from the deformation model are sensitive for unwrapping errors. It can be deduced from Fig. 8.3 that if half a wavelength (28 mm) would be subtracted from the displacements in 1992 and 1993, the solution of a linear displacement rate would have a comparable probability to the current solution. The time gap in 1994 increases the degrees of freedom for the phase unwrapping. Similarly, the period after 2000, in which the temporal sampling drops, is more sensitive to unwrapping errors than the densely sampled period of 1995–2000. Moreover, if the magnitude of subsidence and uplift would have been larger, the estimation of subsidence and uplift based on a linear displacement rate would have been more sensitive for unwrapping errors. The uplift above the Norg field took place during the highest temporal sampling, and hence it has been fully captured by PSI. The Norg displacement time series shows the high potential of PSI for monitoring reservoir behavior, but at the same time it stresses the importance of the a-priori conditions for successful application of PSI (sampling rate, magnitude and

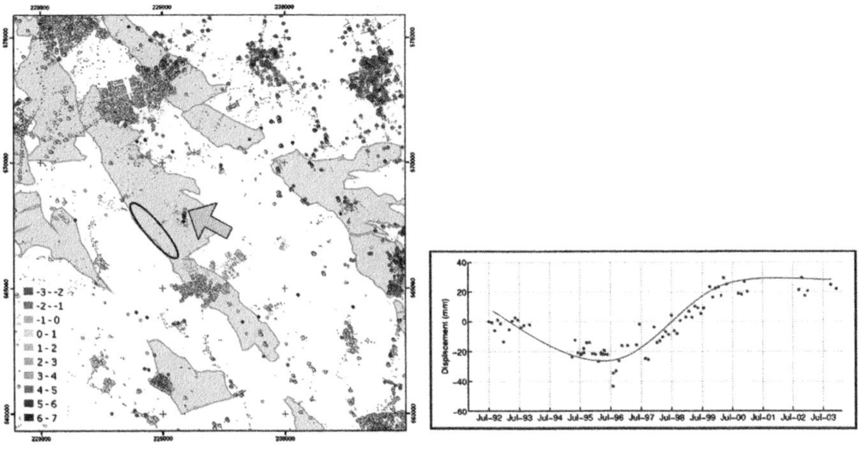

Fig. 8.3 Average displacement rates in mm/year (left) and PS displacement (mm) time series (right) in the Norg area (see Fig. A.1). Subsidence due to gas extraction is followed by uplift due to the use of the reservoir as an underground gas storage from 1997. The area in which the injection wells are located is marked with a black circle. Note that the displacements in 1992–1993 and from 2001 are sensitive for unwrapping errors. The ambiguity in the displacements is equal to half a wavelength, ~28 mm. The deviation of the actual ground movements from the null hypothesis of a linear displacement rate combined with the time gaps, may lead to unwrapping errors. Overlay created in ArcGIS®

extent of subsidence signal). Here, continuous GPS monitoring at a representative location can aid the unwrapping of PSI phase observations. The integration of PSI and GPS is a powerful combination of the spatial observation density of PSI and the higher temporal sampling of GPS.

8.3.2 Spatial Behavior of Subsidence Due to Gas Extraction

Besides the temporal density, the spatial density of PSI is high, especially in urbanized areas (1–2 benchmarks per km^2 for leveling versus ~40 PS/km^2 for PSI). Figure 8.4 and Fig. 8.5 show the PS density in the Waddenzee region for ascending and descending tracks only, whereas Fig. 8.6 shows the combined solution. The delayed onset of gas production in the Anjum field is recognizable in the time series. At the dike locations it is clearly visible that the ascending and descending acquisitions monitor complementary scatterers. The Anjum dike, covered with basalt blocks, is aligned towards the descending viewing geometry. In the ascending track, hardly any scatterers are found on the dike.

8.4 Future Subsidence Monitoring

The high correlation coefficient between leveling and PSI displacement rates (0.94) suggests that PSI is a mature alternative and an independent and reliable comple-

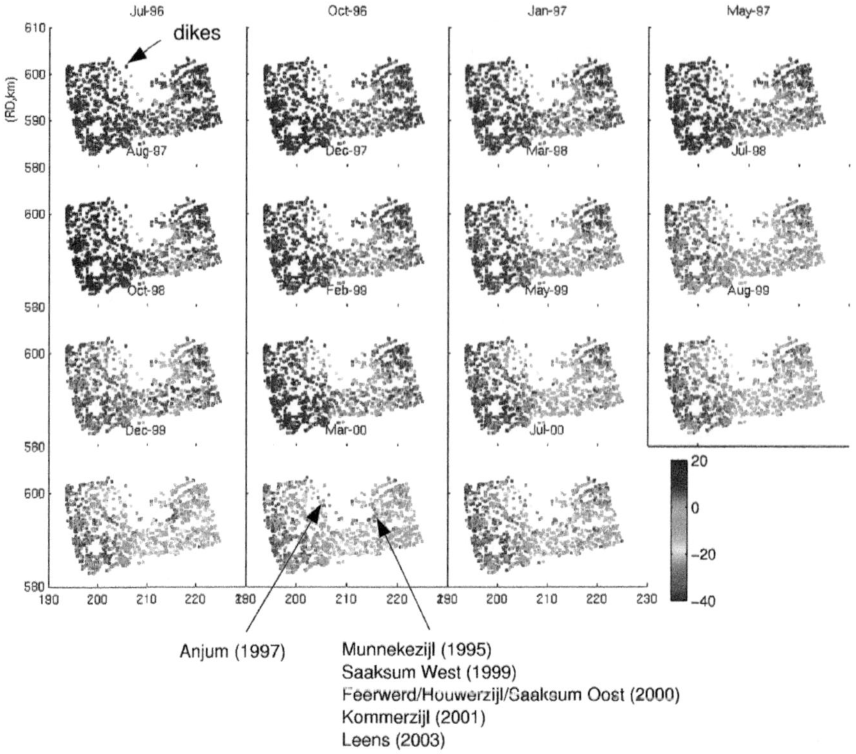

Fig. 8.4 Onshore area near the Waddenzee (see Fig. A.1): PS displacements (mm) estimated from the ascending tracks 487 and 258. An increase in PS displacements is visible after new gas fields have been taken into production. The dikes near Anjum are aligned towards the descending look direction. Hence, no scatterers are found on the dikes in the ascending mode, contrary to Fig. 8.5

mentary technique for future subsidence monitoring in the Groningen region. Temporal updates each 35 days enable more detailed subsidence monitoring than leveling measurements each 2–5 years.

PSI subsidence monitoring is dependent on satellite missions. All satellite missions have a limited lifetime (5–10 years). This means that subsidence above the Groningen gas field, that will be ongoing for at least several decades, needs to be monitored by multiple satellite missions. For a profound reliability assessment, multiple independent overlapping tracks are required. Envisat for example, currently covers only one track at 35 days intervals in image mode. The availability of only one track decreases the reliability, but due to the spatial extent of the scenes (100×100 km), residual errors that propagate over a large spatial extent can still be resolved. Furthermore, the Groningen subsidence pattern is redundantly sampled due its smooth behavior over a large spatial extent.

An option for increasing the reliability of single-track PSI is integration with another (complementary) geodetic measurement technique. The existing leveling network could be reduced, covering only the areas where the PS density is low. An-

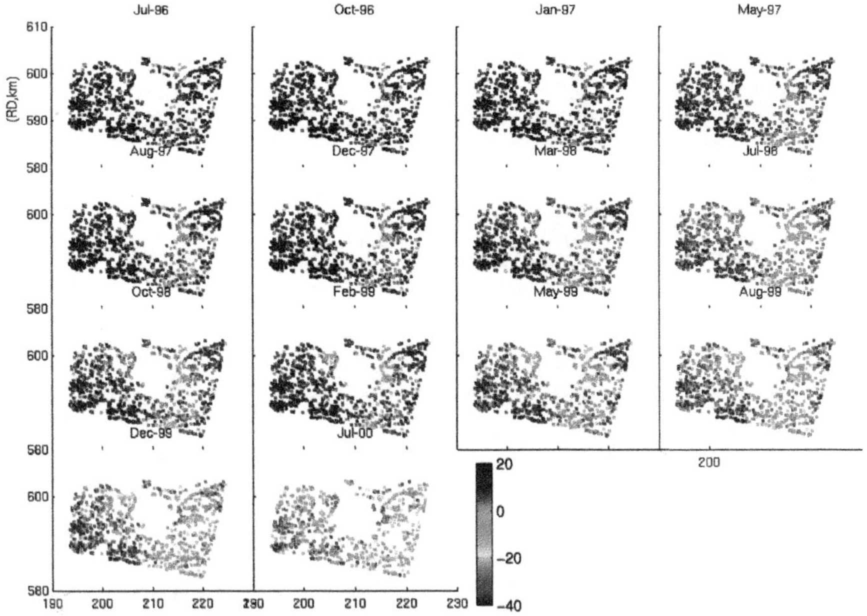

Fig. 8.5 Onshore area near the Waddenzee (see Fig. A.1): PS displacements (mm) estimated from the descending tracks 380 and 151. An increase in PS displacements is visible after new gas fields have been taken into production, similar to Fig. 8.4. In the descending mode, the dikes near Anjum serve as Persistent Scatterers

other option is to deploy several (semi-)continuous GPS monitoring stations, that cover a large spatial extent to resolve residual large-scale errors in PSI. GPS has the advantage that horizontal components are measured as well, which contain useful information regarding reservoir behavior. When establishing a sparse network of GPS stations, the foundation with reference to the Pleistocene layer should be guaranteed, to avoid additional uncertainty on the deformation regime that is monitored.

It is often suggested to establish a network of corner reflector for PSI monitoring, especially in areas that are poorly covered. Corner reflectors can be useful, but their effect should not be overestimated. When placing corner reflectors, a new time series has to be built up. Furthermore, they have to be carefully mounted (on the Pleistocene layer, or at structures such as bridges that can be incorporated in the leveling network) and maintained. When explicitly placing corner reflectors, one could also consider GPS stations, with an even higher temporal sampling and moreover providing horizontal displacements as well. Corner reflectors have to be precisely aligned towards the satellite, requiring advanced procedures or different reflectors for multiple satellites and tracks. Establishing one or several corner reflectors can be beneficial for the following reasons:

1. establishing a common reference for PSI and other geodetic techniques; however this does not imply that measurements referring to other benchmarks and PS can

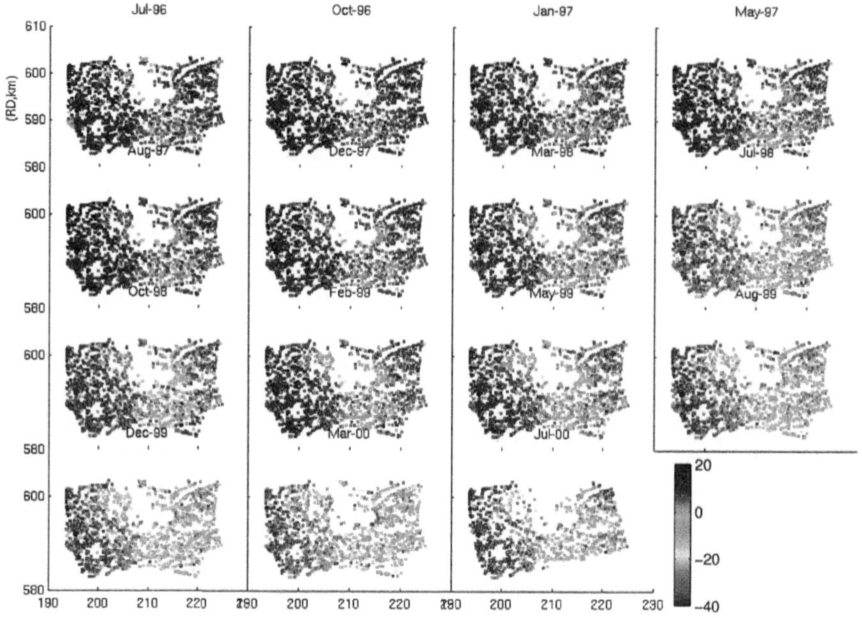

Fig. 8.6 Onshore area near the Waddenzee (see Fig. A.1): combined PS displacements (mm) from both the ascending and descending tracks (487, 258, 380 and 151)

be directly compared since they may represent different deformation regimes, and

2. for the connection to absolute subsidence measurements (gravity measurements, GPS).

The strength of PSI as a subsidence monitoring technique lies in the availability of natural targets that act as Persistent Scatterers. The optimal way to integrate PSI with other geodetic techniques would be by establishing pole type reflectors that are monitored from all tracks. These pole type reflectors may already exist as natural scatterers, e.g., wind mills.

Finally, the importance of a backup procedure has to be stressed. Since satellite missions have a limited lifetime and failures may occur, it should always be possible to restart the leveling campaigns. Therefore, it is essential to maintain all existing benchmarks.

Chapter 9
Conclusions and Recommendations

In this chapter, the conclusions are summarized. Additionally, the research contributions are listed. To conclude, recommendations for future research are given.

9.1 Conclusions

It can be concluded that the PSI technique can be applied for monitoring subsidence due to gas extraction in the Groningen area, provided that the quality measures are clearly defined. These quality measures comprise the precision and reliability of the PSI technique itself and the idealization precision for the estimation of subsidence due to hydrocarbon production (Chap. 4). The precision of PSI deformation estimates follows from the PSI estimation procedure, and has to be accompanied by an assessment of the influence of possible model errors on these estimates (Sect. 4.2). The reliability of the PSI estimates are quantified by the misclosures after the datum connection of independent overlapping tracks that observe the same deformation signal (Chap. 5). Moreover, the PS selection method to improve the idealization precision has to be stated (selection method based on spatial correlation or PS characterization tools).

The PSI results of both ERS and Envisat clearly depict areas affected by surface movements in the entire northern part of the Netherlands, which correspond with the gas production areas (Sect. 6.2). The total subsiding area above the Groningen gas field has a diameter of ~50 km. Relative subsidence rates up to ~7 mm/year in the period 1992–2007 can be deduced. The subsidence pattern is spatially coherent. Since the precision and reliability of the PSI deformation estimates have been assessed independent of other measurement techniques, PSI is considered as a stand-alone technique for deformation monitoring.

Cross-validation with the leveling displacement rates reveals a correlation coefficient of 0.94 (Sect. 7.2), similar to the correlation coefficient of leveling and PSI displacements in the controlled corner reflector experiment, 0.94 (Sect. 4.4). Moreover, the correlation coefficient of displacement estimates from repeated leveling campaigns is also lower than 1 (~0.94–0.97), due to the measurement precision. Using the leveling results as a reference, this means that PSI has reached the maturity to be operationally used for monitoring subsidence due to gas extraction in the northern part of the Netherlands—stand-alone or, in specific cases, in concert with significantly reduced leveling campaigns or GPS.

The central problem statement has been defined as:

Is the InSAR technique able to provide precise and reliable deformation estimates for the monitoring of subsidence due to hydrocarbon production in the Netherlands, particularly in the Groningen region?

V.B.H. (Gini) Ketelaar, *Satellite Radar Interferometry*,
Remote Sensing and Digital Image Processing 14,

It has been divided into the following sub-questions that will be subsequently answered.

1. Does the area of interest contain sufficient radar targets with coherent phase observations?
2. Does InSAR provide a precise estimation of surface displacements in the Groningen area?
3. How can we assess the reliability of InSAR deformation estimates?
4. Is it possible to estimate subsidence due to hydrocarbon production from InSAR measurements in the presence of multiple deformation phenomena?
5. Are the PSI deformation estimates in agreement with the leveling results?
6. Can InSAR aid the understanding of reservoir behavior?
7. Is the continuity of subsidence monitoring using InSAR guaranteed?

In the following sections we will conclude on these questions.

9.1.1 PS Density

The density of the accepted PS in the second order network varies from 0–10 PS per km^2 in rural areas to more than 100 PS per km^2 in the urbanized areas (Sect. 6.2.1). Approximately 80% of the Groningen subsidence area is covered with more than one PS per km^2 using a single satellite track, with an average density of ~40 PS per km^2. For comparison, following the guidelines of Duquesnoy (2002), the leveling benchmark density for subsidence monitoring of the Groningen gas field is 1–1.5 benchmark per km^2, to capture the spatial subsidence pattern. Both the leveling and PSI measurement points are located at buildings and other man-made objects, mainly located along the existing infrastructure. Although local agricultural areas may not be covered with measurement points, both leveling and PSI meet the point density guidelines.

9.1.2 Precision

Precision is defined as the dispersion of the deformation estimates around their expectation value, and is represented by the variance-covariance matrix. It is a superposition of observational noise, and noise due to unmodeled deformation, imprecise APS estimates, processing induced errors and unmodeled systematic errors (sub-pixel position, residual orbital errors). Moreover, the precision of the PSI deformation estimates depends on the acquisition geometry in space and time, and the PS density, especially in the first order network. Here, we conclude that:

- The relative precision (1-sigma) of *displacement rates* depends on the number of acquisitions and the distance between the PS. For the smallest Groningen stack consisting of 24 interferograms the relative precision is ~0.1 mm/year per

$\sqrt{\text{km}}$; for the largest stack of 74 interferograms it is ~0.04 mm/year per $\sqrt{\text{km}}$ (Sect. 6.3).

- The precision (1-sigma) of double-difference *displacements* is on average 3 mm, with respect to the closest accepted PS in the first order network. Dependent on the physical properties of the PS and the relative arc length, it is ≤3 mm in urban areas, where PS are generally less than 500 m apart. In rural areas, where the distance between PS can exceed one kilometer, the precision of displacements is 3–7 mm (Sect. 6.3).

9.1.3 Reliability

Reliability is defined as the sensitivity for, and detectability of, model imperfections. Reliability can only be assessed in case of redundancy. The PSI system of observations equations that includes the unknown ambiguities contains more unknowns than observations. There is no redundancy, even a rank deficiency, that needs to be resolved by introducing pseudo-observations. Testing on model imperfections is therefore not possible in the estimation of deformation between two PS.

However, the effect of unmodeled phenomena on the parameter estimates has been assessed under the assumption of ambiguity resolution success rates of 1 (Sect. 4.2). The model errors comprise sub-pixel position and orbit errors. Simulations have been performed, using the acquisition geometry of the six Groningen ERS tracks. This has resulted in the following conclusions:

- The azimuth sub-pixel position has not been taken into account during the PSI estimation. This can lead to a maximum additional error of ~0.5 mm/year. This error is independent of the mutual PS distance and depends on the relative sub-pixel error and the Doppler centroid frequency difference in time.
- Random (residual) orbital errors of 5 and 8 centimeters in radial and across-track direction can lead to a PS velocity error of ~1 mm/year over 100 km. This error has a systematic character that propagates over a large spatial extent.

Since the ambiguity resolution success rates have been assumed 1, this is the most optimistic scenario for the effect of model imperfections. Simulations in Sect. 4.2.4 have shown that unwrapping success rates of 1 can not be guaranteed in rural areas with a low PS density, i.e. the probability of correct unwrapping is not 100%. For a strict reliability assessment, redundancy is required, which can be obtained by the use of independent overlapping tracks.

For the Groningen subsidence area, six overlapping tracks provide independent observations of the same subsidence signal. Model errors can be detected through the integration of these tracks by means of a datum connection procedure (Sect. 6.4). This datum connection consists of two steps, i.e. (1) conversion to the radar coordinate system of the master track, and (2) connection of the PSI parameter estimates (displacements, heights).

The quality of the datum connection and the reliability of the PSI deformation estimates have been evaluated by means of multi-track PS clusters. The standard deviation of the displacement rates in 70% of the PS clusters is less than 1 mm/year after datum connection. Although the PSI results of all tracks have been merged into a consistent system defined by the master track, the presence of a small spatial trend of several mm/year over a 100 km distance has to be taken into account. Due to the wide coverage of a single scene, this trend can be corrected for, provided that the majority of the scene is not affected by surface deformation.

PSI deformation estimates are in the satellite line of sight direction. Neglecting the horizontal displacement component can lead to errors up to ~1 mm/year in the vertical displacement rates. Multiple viewing geometries (ascending, descending, and adjacent) can resolve different displacement components. This has resulted in horizontal displacement rates of ~2–3 mm/year towards the center of the subsidence bowls, observed for the first time. The magnitude and direction of the horizontal displacement rates match with the theoretically predicted.

9.1.4 Deformation Regimes

The InSAR deformation estimates are not necessarily related to the deformation signal of interest. Remote sensing techniques such as InSAR monitor all surface displacements from space, regardless of the deformation mechanism. The presence of different deformation regimes (structural instabilities, shallow and deep mass displacements) requires an additional interpretation step for the estimation of the signal of interest (Sect. 4.5). The interpretation of deformation estimates is not unique for InSAR: leveling benchmarks that are not properly founded on a stable subsurface layer can exhibit autonomous movements as well.

Compared to traditional geodetic techniques, the physical measurement point is less well defined for InSAR: it has a lower idealization precision for deformation monitoring. Different methodologies have been investigated to improve PS characterization (Sect. 6.5). Eventually, the differences in the PS velocity histograms before and after selection have appeared not to be significant (<0.5 mm/year), in two case study areas. This implies that the PS reflections (direct and indirect) in these case study areas refer to buildings and structures that are subject to the same deformation regime.

Since subsidence due to gas extraction is the common deformation regime, PS can be selected (and filtered) based on spatial correlation. If the majority of the PS refer to well-founded buildings, the peak of the velocity histogram refers to the displacement rate due to gas extraction. However, this does not hold in areas were the majority of the buildings and structures have a bad foundation (instable subsurface layers, peat oxidation in combination with shallow foundations). This may be the cause for the differences between the leveling and PSI displacement rates in the southeastern part of the Groningen subsidence bowl, see Fig. 7.12. Here, manual selection of PS targets, or additional measurements from different geodetic techniques referring to the stable Pleistocene layer, may be locally required.

Besides PS selection for the estimation of subsidence due to hydrocarbon production, the displacement decomposition due to different deformation regimes has been investigated. This method has the advantage that it uses all PS displacements, and is performed by variance component estimation. Its application is however limited, because spatio-temporal covariance functions of different deformation regimes need to be independent and high redundancy is required to be able to estimate the stochastic model parameters with a reasonable precision.

9.1.5 Cross-Validation PSI and Leveling

The historical leveling results have been compared with the PSI deformation estimates, which have been converted to displacement rates along the vertical. The velocity differences have a standard deviation less than 1 mm/year. The displacement rates of leveling benchmarks and the mean of neighboring PS have a correlation coefficient of 0.94 (Sect. 7.2). Considering that the correlation coefficient of displacement estimates from repeated leveling campaigns is also lower than 1 ($\sim$0.94–0.97), due to the measurement precision, the correlation between PSI and leveling displacement estimates approaches its maximum.

It has been shown that the spatio-temporal observation density of PSI can overcome a lower observational precision ($\sim$3 mm for displacement estimates between two PS) compared to the precision of leveling height difference measurements (1 mm/$\sqrt{\text{km}}$). A stack of $\sim$25 SAR acquisitions is required to obtain the same precision of displacement rates (Sect. 7.1). A larger stack however will even provide displacement rates of a higher precision than the leveling technique. The spatial sampling of PSI increases the precision of the deformation estimates as well. However, due to the correlation between PSI double-difference observations, the increase in precision is lower than for uncorrelated deformation estimates.

A framework for the integration of observations from multiple geodetic techniques can be defined using the Best Linear Unbiased Prediction methodology (Sect. 7.3). Moreover, this framework can be utilized for the integration of displacement estimates from different sensors, with a possible time gap. However, the application in PSI is complicated if ambiguity resolution success rates of 1 are not obtained. In this situation, the probability density function of the PSI estimates is multi-modal. The evaluation of alternative hypotheses and the definition and interpretation of precision and reliability measures will not be straightforward, moreover due to the uncertainties in the stochastic model of PSI.

9.1.6 Hydrocarbon Reservoir Behavior

The spatial and temporal sampling of InSAR, and the spatial coverage of an interferogram is significantly higher than conventional monitoring techniques, such as leveling, can provide. Therefore, the PSI deformation estimates contain information about the behavior of the hydrocarbon reservoirs that could not be extracted in the past (Sect. 8.3).

The subsidence delay is one of the reservoir characteristics that has gained attention during the preparations for the production below the Waddenzee (NAM, 2006). Due to the high temporal sampling rate (up to four acquisitions each 35 days versus one every 2–5 years for leveling), the subsidence delay can be determined more precisely. For the Anjum gas field, subsidence starts several months to one year after the start of gas extraction. Another feature that has been captured by PSI is the uplift due to underground gas storage in Norg. An uplift of four centimeters in four years has been observed after start of the gas storage in 1997.

9.1.7 Outlook

The period that subsidence monitoring in Groningen is required (in the order of decades) exceeds the lifetime of a satellite (5–10 years). Continuity in the time series has been shown in the displacement time series of one track between ERS and Envisat, see Sect. 6.2.2. The future outlook of satellite missions is positive (TerraSAR-X, Sentinel-1, ALOS, RADARSAT-2, Cosmo-SkyMed). Each new satellite mission will have its own characteristics, and needs to be carefully validated. The integration of deformation estimates of multiple missions has to be performed in the parameter space, as the individual measurements cannot be related unambiguously. Since the displacement rates due to gas extraction are near-linear in time, the displacement observations can be easily connected. To ensure continuity in subsidence monitoring, the leveling benchmarks have to be maintained. They serve as a backup if a satellite fails or a mission is not continued.

9.2 Contributions

Summarizing, this research has made the following contributions:

1. Pseudo-calibration has been introduced in the PS selection procedure, which saves computation time and data storage space.
2. It has been demonstrated for the Groningen region that the PS density in rural areas is sufficient for a precise and reliable estimation of subsidence due to gas extraction.
3. The influence of model errors (sub-pixel positions, orbit errors, unwrapping errors, sidelobe observations) on the deformation estimates has been quantified.
4. It has been shown that a reliability assessment of PSI can be performed by the datum connection of multiple, overlapping, independent satellite tracks.
5. The applicability of PSI for monitoring deformation that is characterized by small displacement rates (up to ~ -7 mm/year), over a large spatial extent ($>$200 km) has been demonstrated.

6. The different viewing geometry of adjacent and cross-heading tracks has been utilized to obtain horizontal displacement rates for the Groningen gas field.
7. It has been shown that the correlation coefficient of leveling and PSI displacement rates (0.94) is comparable to the correlation of displacements that have been obtained in the controlled corner reflector experiment (0.94).
8. The concept of idealization precision for PSI deformation monitoring in the presence of multiple deformation regimes has been introduced.
9. The application of variance component estimation for the separation of autonomous movements and spatially correlated subsidence signal has revealed ground movements that could be physically explained (uplift due to the water injection in the Rotterdam area).
10. It has been shown that PS characterization is not necessarily required for the estimation of subsidence due to gas extraction (PS selection based on spatial correlation is sufficient in areas where the majority of the buildings is well-founded).
11. The application of PSI for monitoring reservoir behavior in time has been demonstrated (subsidence delay, uplift due to underground gas storage).

9.3 Recommendations

Although the PSI displacement rates of the ERS and Envisat tracks clearly capture surface movements due to gas extraction, and the correlation with the leveling technique is satisfactory, it is recommended to investigate if the PSI processing for large (rural) areas can be further optimized.

1. The estimation of error sources, such as atmospheric signal, is dependent on the density and quality of PS candidates in the first order network. In the current Groningen results, observations from spatially non-fitting arcs are removed from all interferometric combinations to avoid the acceptance of erroneous ambiguity resolutions. The disadvantage is that the resulting first order network is very sparse. Coherence thresholds to determine the set of accepted PS should be finetuned, taking the false detection rate as a function of the number of images into account. The PS density should be optimized by selection of PS based on a-priori knowledge on the terrain characteristics (location of buildings and structures). New sensors, such as TerraSAR-X and Cosmo-SkyMed, that have a higher resolution, can potentially contribute to a higher PS density as well. Furthermore, the quality of the individual interferograms should be integrated in the estimation procedure.
2. The maximum unwrapping success rates that can be obtained have to be investigated. This can be done by using small 3D, in space and time, (spider) networks that are recursively connected. Using 3D networks also has the advantage that the stochastic model can be improved by variance component estimation of both spatial and temporal parameters.
3. To limit biases in the velocity estimates in the presence of high Doppler centroid frequency deviations, the azimuth sub-pixel position has to be incorporated in the system of equations.

4. The estimation of orbit errors should be improved to be able to unambiguously estimate ground movements of low magnitude (several mm/year or less) over a large spatial extent (>100 km).
5. Continuous time series from multiple sensors and independent overlapping tracks are essential for continuous and reliable subsidence monitoring. The time interval between subsequent acquisitions should be $\sim$35 days or less; the resolution needs to be at ERS/Envisat level or better; and the phase observations need to result in a precision of displacement estimates of $\sim$3 mm or better. For the reliability assessment, it is recommended that the subsidence signal is monitored by at least one ascending and one descending track. For future acquisitions this should be secured. Multiple tracks are also needed for the decomposition into horizontal and vertical displacements.
6. For future subsidence monitoring, regular updates of the displacement time series are required. This can be obtained by dynamic PSI processing (Marinkovic and Hanssen, 2007), which provides recursive updates and hence is more efficient than the recalculation of batch solutions.
7. Alternating Polarization data needs to be further investigated in areas that are known for shallow subsurface movements combined with shallow foundations of buildings. Only one or several acquisitions are required for interpretation purposes; the PSI time series should be minimally interrupted.
8. Considering the interpretation of PSI deformation estimates, the validity of the application of statistical methods (correlation length) has to be carefully judged in the area of interest. The deformation signal of interest has to be dominant in the majority of the PS displacements, which does not hold in areas where most of the buildings and structures have instable foundations in combination with soft shallow soils.
9. It is recommended to further investigate the potential additional information in PSI deformation estimates. The temporal sampling frequency has proven to provide additional insight in reservoir behavior. The spatial sampling frequency and extent has to be further analyzed with respect to reservoir behavior. The contribution of new satellite missions, such as TerraSAR-X and Cosmo-SkyMed, with a higher resolution, a higher repeat interval and the ability to acquire fully polarimetric data, should certainly be investigated.

Considering the near future, subsidence monitoring in the Groningen region using Envisat has to be performed in a controlled way since only one track has been sufficiently monitored in image mode up to 2008 ($\sim$40 acquisitions). Although it is possible to perform a (limited) reliability assessment due to the redundant spatial sampling of the deformation signal and the large spatial coverage (to control systematic errors that propagate over a large spatial extent), multiple tracks enable a much stronger evaluation of the reliability.

Although new satellites have been launched or are in preparation, it is essential that the leveling benchmarks are maintained. Since satellite missions have a limited lifetime and failures may occur, it should always be possible to perform a leveling campaign, when required.

Appendix 1
Location of Research Areas

V.B.H. (Gini) Ketelaar, *Satellite Radar Interferometry*,
Remote Sensing and Digital Image Processing 14,

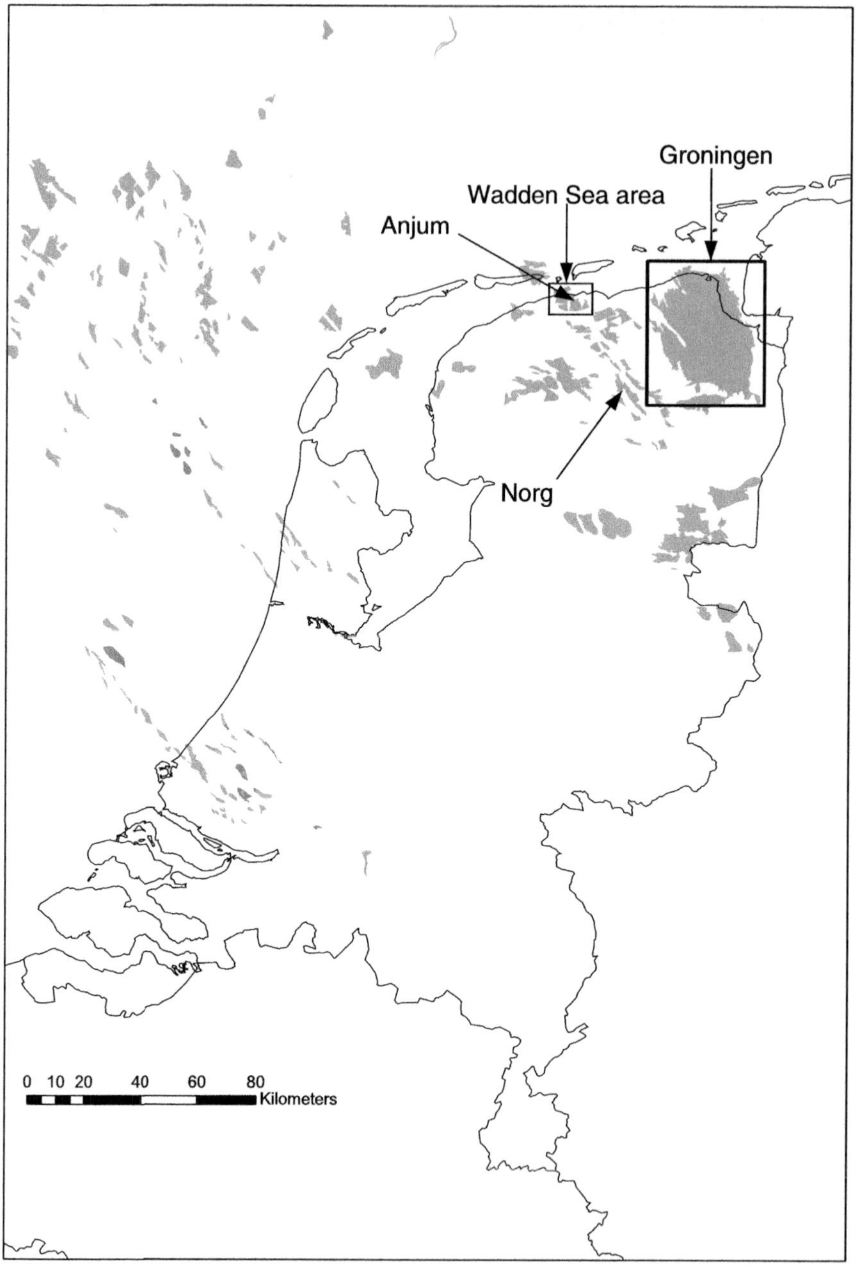

Fig. A.1 Location of Groningen, Anjum, and Norg gas fields, and indication of the gas fields in the Waddenzee area (onshore and offshore)

Appendix 2
PSI and Leveling Displacement Profiles

A2.1 PSI (Track 380,487) and Leveling (Free Network Adjustments)

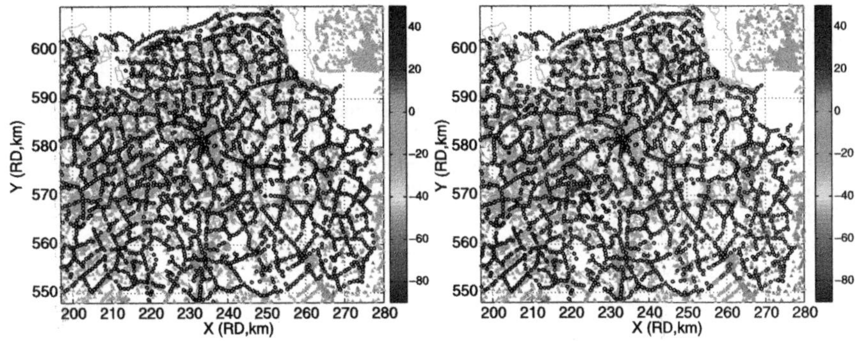

Fig. A.2 Estimated displacements (mm) in the periods 1993–1998 (left) and 1993–2003 (right) for PSI (triangles) and leveling (circles). The leveling displacement estimates have been computed from the benchmark heights obtained by free network adjustments of subsequent epochs. The PSI displacement estimates are from the tracks 380 and 487. The boundaries of the gas fields are depicted in gray

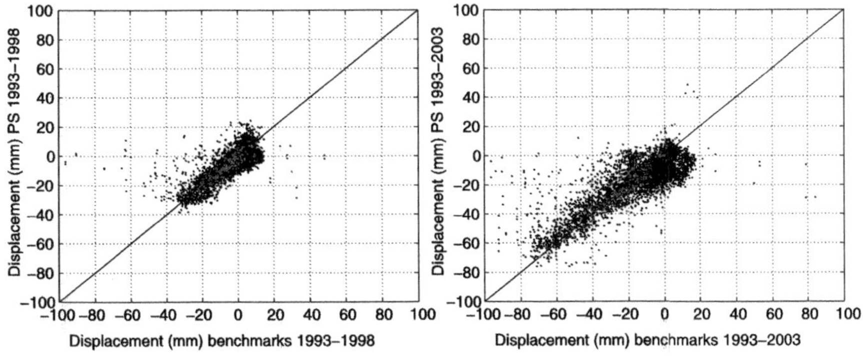

Fig. A.3 Correlation between PSI and leveling displacements (mm) in the periods 1993–1998 (left) and 1993–2003 (right). The leveling displacement estimates have been computed from the benchmark heights obtained by free network adjustments of subsequent epochs. The PSI displacement estimates are from the tracks 380 and 487. The red dots depict the selection of locations of stable benchmarks (Schoustra 2006). The correlation before and after selection of stable benchmark locations is 0.74 and 0.87 respectively in the period 1993–1998. In the period 1993–2003 the correlation before and after selection of stable benchmark locations is 0.81 and 0.94 respectively

V.B.H. (Gini) Ketelaar, *Satellite Radar Interferometry*,
Remote Sensing and Digital Image Processing 14,

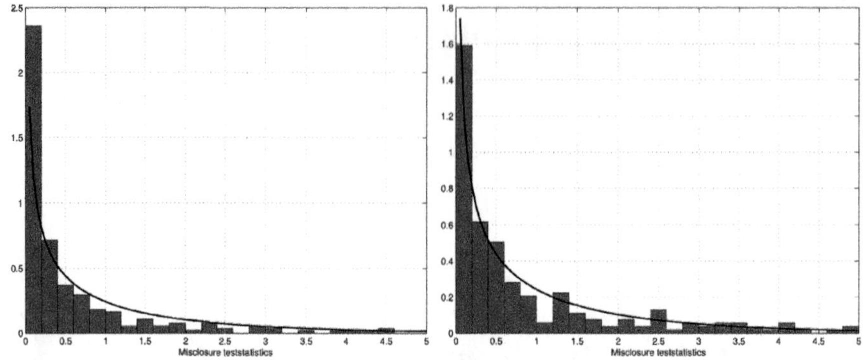

Fig. A.4 PSI and leveling misclosure teststatistics and their theoretical χ^2 distribution between PSI and leveling displacements (mm) in the periods 1993–1998 (left) and 1993–2003 (right). The leveling displacement estimates have been computed from the benchmark heights obtained by free network adjustments of subsequent epochs. The PSI displacement estimates are from the tracks 380 and 487

A2.2 PSI (Track 380,487) and Leveling (SuMo Analysis)

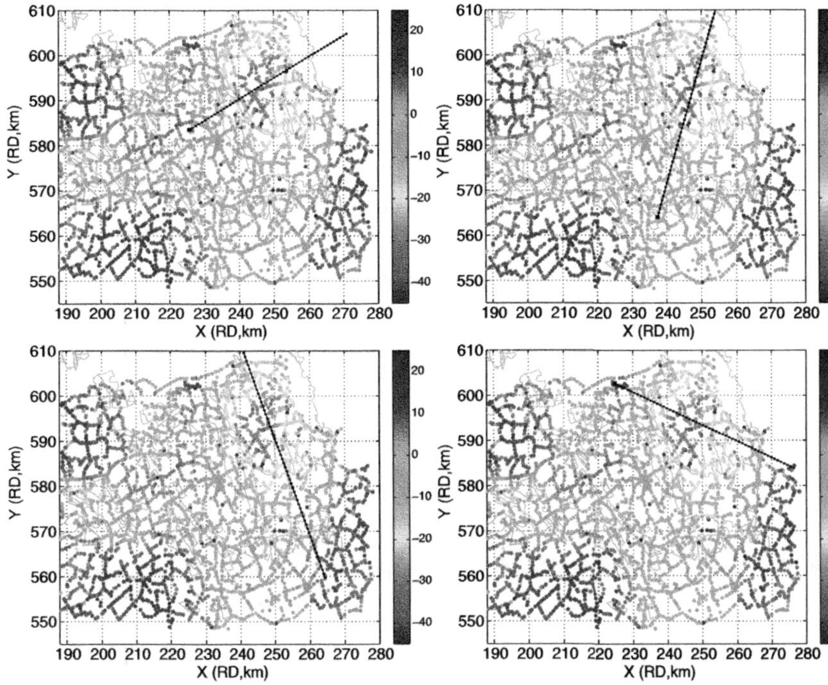

Fig. A.5 Four profiles that depict the PSI and leveling evaluation locations. The displacements at each evaluation location have been computed as the weighted average of all displacement estimates within a radius of one kilometer. The profiles 1, 2, 3 and 4 are the top-left, top-right, bottom-left and bottom-right profile respectively. The depicted displacement estimates are the leveling displacements estimates (mm) in the period 1993–1998, based on the free network adjustments the 1993 and the 1998 epochs

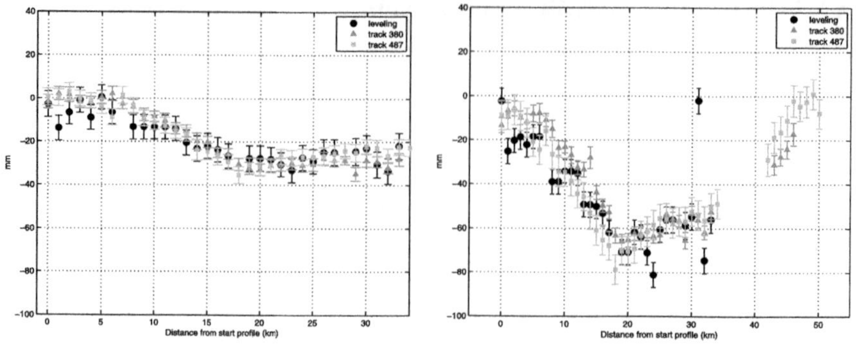

Fig. A.6 PSI and leveling displacement estimates (mm) at the evaluation locations along profile 1 for the periods 1993–1998 (left) and 1993–2003 (right). The leveling displacement estimates have been computed from the benchmark heights obtained by free network adjustments of subsequent epochs. The PSI displacement estimates are from the tracks 380 and 487

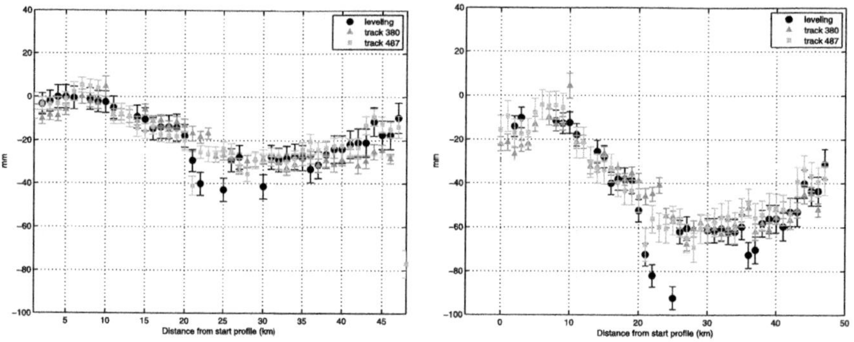

Fig. A.7 PSI and leveling displacement estimates (mm) at the evaluation locations along profile 2 for the periods 1993–1998 (left) and 1993–2003 (right). The leveling displacement estimates have been computed from the benchmark heights obtained by free network adjustments of subsequent epochs. The PSI displacement estimates are from the tracks 380 and 487

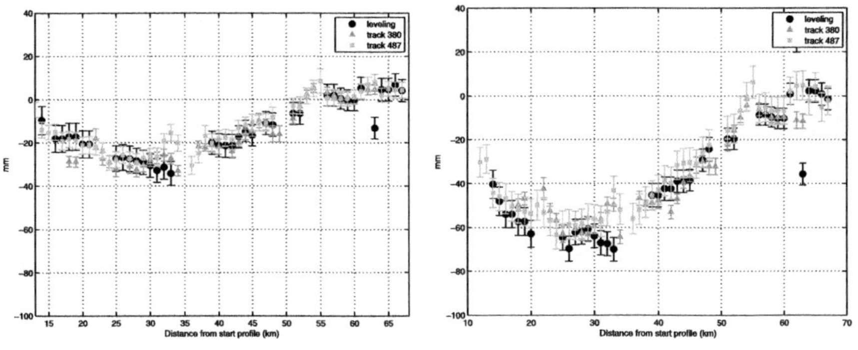

Fig. A.8 PSI and leveling displacement estimates (mm) at the evaluation locations along profile 3 for the periods 1993–1998 (left) and 1993–2003 (right). The leveling displacement estimates have been computed from the benchmark heights obtained by free network adjustments of subsequent epochs. The PSI displacement estimates are from the tracks 380 and 487

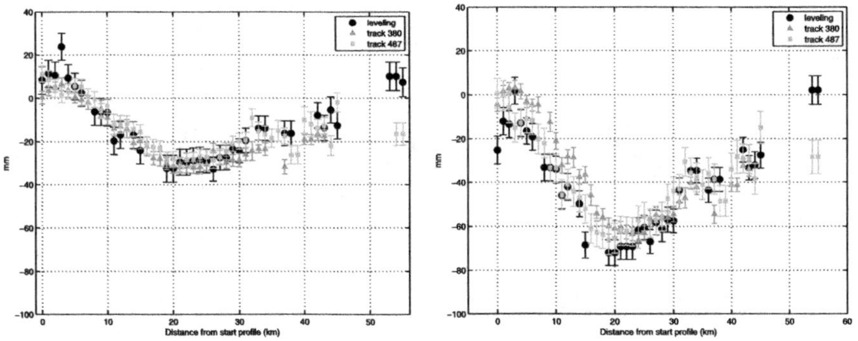

Fig. A.9 PSI and leveling displacement estimates (mm) at the evaluation locations along profile 4 for the periods 1993–1998 (left) and 1993–2003 (right). The leveling displacement estimates have been computed from the benchmark heights obtained by free network adjustments of subsequent epochs. The PSI displacement estimates are from the tracks 380 and 487

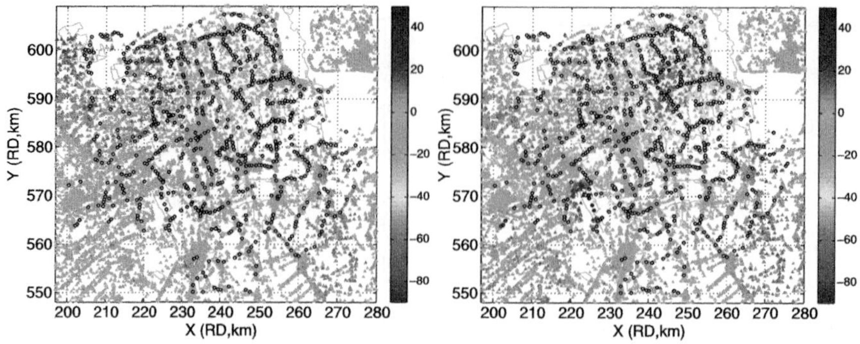

Fig. A.10 Estimated displacements (mm) in the periods 1993–1998 (left) and 1993–2003 (right) for PSI (triangles) and leveling (circles). The leveling displacement estimates have been obtained from the SuMo 2003 analysis (Schoustra 2004). The PSI displacement estimates are from the tracks 380 and 487. The boundaries of the gas fields are depicted in gray

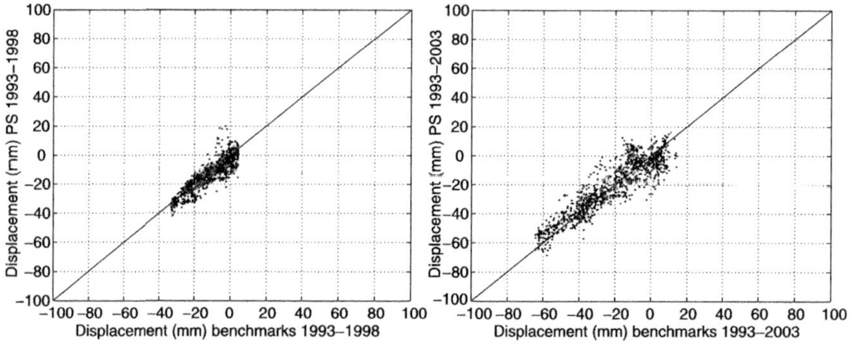

Fig. A.11 Correlation between PSI and leveling displacements (mm) in the periods 1993–1998 (left) and 1993–2003 (right). The leveling displacement estimates have been obtained from the SuMo 2003 analysis (Schoustra 2004). The PSI displacement estimates are from the tracks 380 and 487. The red dots depict the selection of locations of stable benchmarks (Schoustra 2006). The correlation before and after selection of stable benchmark locations is 0.90 and 0.93 respectively in the period 1993–1998. In the period 1993–2003 the correlation before and after selection of stable benchmark locations is 0.91 and 0.95 respectively

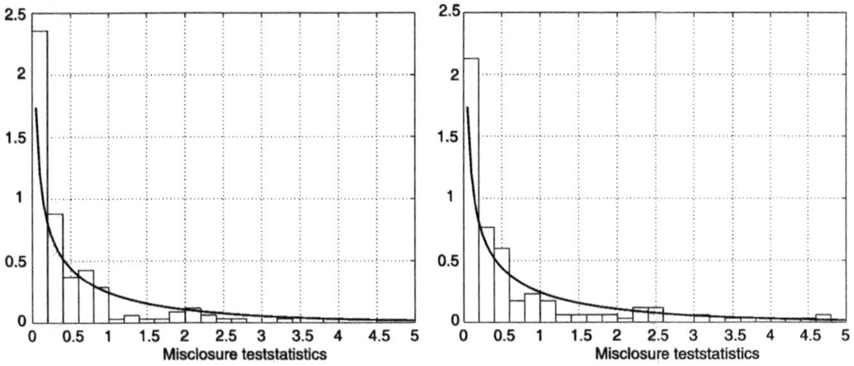

Fig. A.12 PSI and leveling misclosure teststatistics and their theoretical χ^2 distribution between PSI and leveling displacements (mm) in the periods 1993–1998 (left) and 1993–2003 (right). The leveling displacement estimates have been obtained from the SuMo 2003 analysis (Schoustra 2004). The PSI displacement estimates are from the tracks 380 and 487

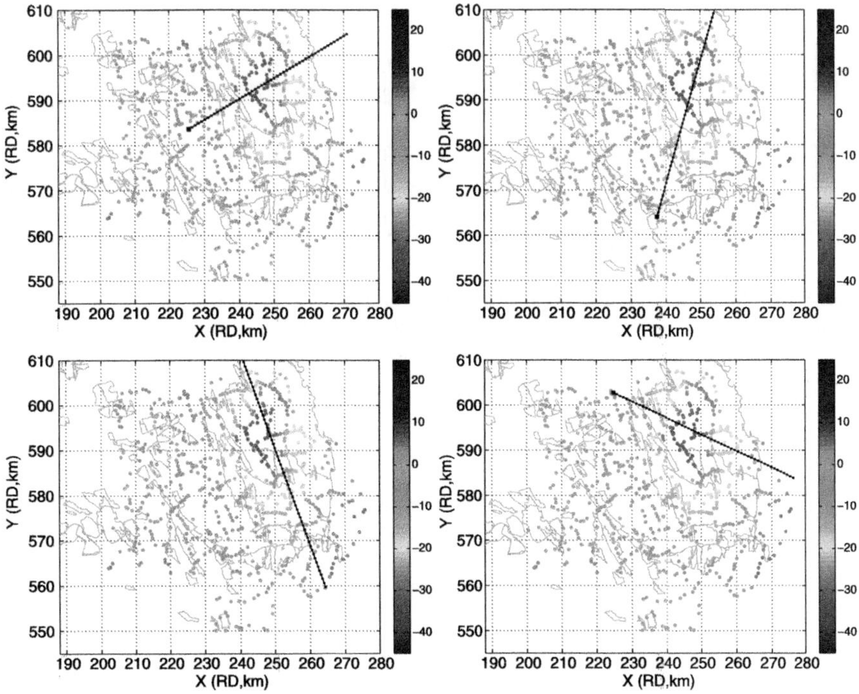

Fig. A.13 Four profiles that depict the PSI and leveling evaluation locations. The displacements at each evaluation location have been computed as the weighted average of all displacement estimates within a radius of one kilometer. The profiles 1, 2, 3 and 4 are the top-left, top-right, bottom-left and bottom-right profile respectively. The depicted displacement estimates are the leveling displacements estimates (mm) in the period 1993–1998, based on the SuMo 2003 analysis (Schoustra 2004)

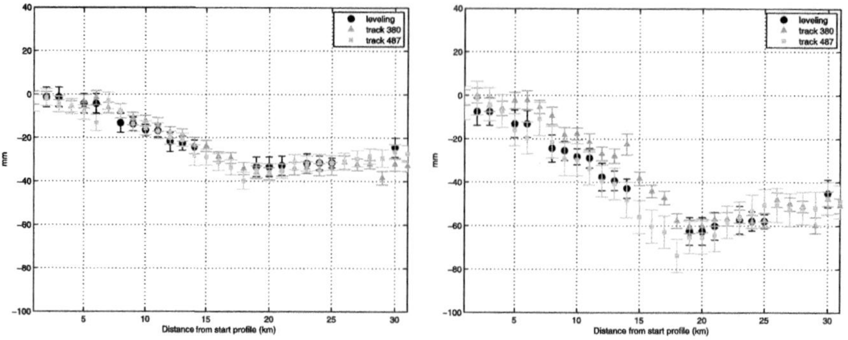

Fig. A.14 PSI and leveling displacement estimates (mm) at the evaluation locations along profile 1 for the periods 1993–1998 (left) and 1993–2003 (right). The leveling displacement estimates have been obtained from the SuMo 2003 analysis (Schoustra 2004). The PSI displacement estimates are from the tracks 380 and 487

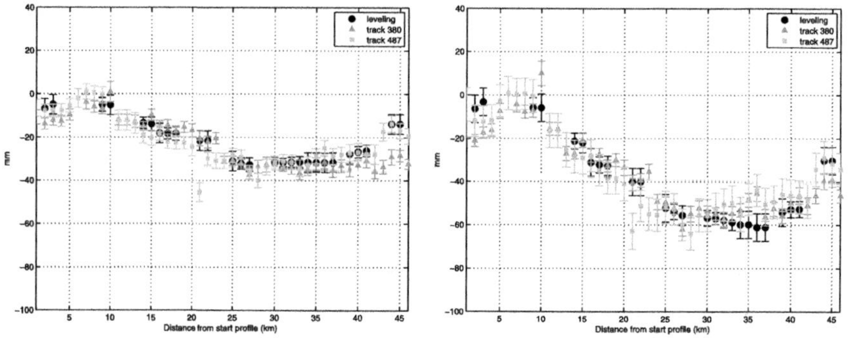

Fig. A.15 PSI and leveling displacement estimates (mm) at the evaluation locations along profile 2 for the periods 1993–1998 (left) and 1993–2003 (right). The leveling displacement estimates have been obtained from the SuMo 2003 analysis (Schoustra 2004). The PSI displacement estimates are from the tracks 380 and 487

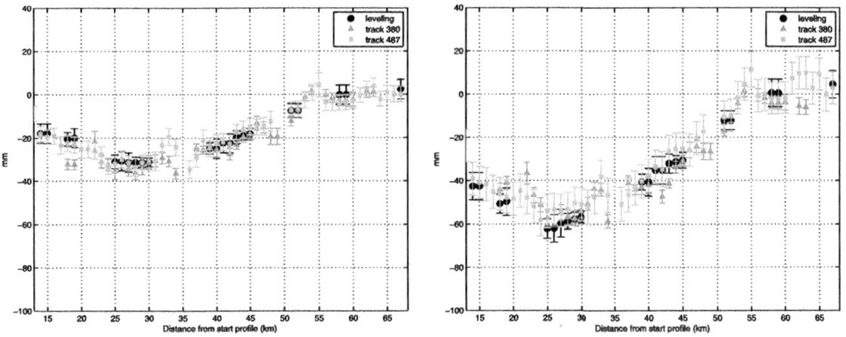

Fig. A.16 PSI and leveling displacement estimates (mm) at the evaluation locations along profile 3 for the periods 1993–1998 (left) and 1993–2003 (right). The leveling displacement estimates have been obtained from the SuMo 2003 analysis (Schoustra 2004). The PSI displacement estimates are from the tracks 380 and 487

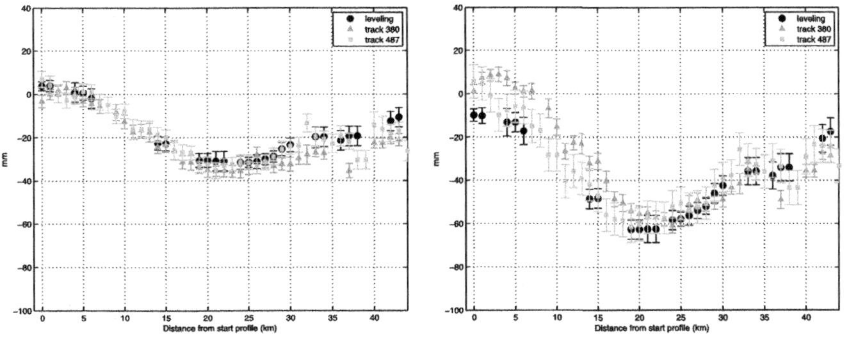

Fig. A.17 PSI and leveling displacement estimates (mm) at the evaluation locations along profile 4 for the periods 1993–1998 (left) and 1993–2003 (right). The leveling displacement estimates have been obtained from the SuMo 2003 analysis (Schoustra 2004). The PSI displacement estimates are from the tracks 380 and 487

Bibliography

Adam, N., Kampes, B. M. and Eineder, M. (2004), The development of a scientific Persistent Scatterer System: Modifications for mixed ERS/ENVISAT time series, in: *ENVISAT & ERS Symposium, Salzburg, Austria, 6–10 September, 2004.*

AGI (2005), Specificaties doorgaande waterpassing, Tech. rep., Adviesdienst Geo-informatie en ICT, Rijkswaterstaat, in Dutch.

AHN (2008), website AHN: Actual Height model of the Netherlands, www.ahn.nl.

Amiri-Simkooei, A. (2007), *Least-Squares Variance Component Estimation; Theory and GPS Applications*, Ph.D. thesis, Delft University of Technology.

Anderson, E. M. (1936), The dynamics of the formation of cone-sheets, ring-dykes, and caldron subsidence, *Proceedings of the Royal Society of Edinburgh*, **56**:128–157.

Baarda, W. (1981), *S-Transformations and Criterion Matrices*, vol. 5 of *Publications on Geodesy, New Series*, Netherlands Geodetic Commission, Delft, 2nd edn.

Berardino, P., Fornaro, G., Lanari, R. and Sansosti, E. (2002), A new algorithm for surface deformation monitoring based on small baseline differential SAR interferograms, *IEEE Transactions on Geoscience and Remote Sensing*, **40**(11):2375–2383.

Bovenga, F., Refice, A., Nutricato, R., Pasquariello, G. and DeCarolis, G. (2002), Automated calibration of multi-temporal ERS SAR data, in: *International Geoscience and Remote Sensing Symposium, Toronto, Canada, 24–28 June 2002.*

Brand, G. B. M. (2002), De historische data van de primaire waterpassingen van het NAP, Tech. rep., Ministerie van Verkeer en Waterstaat, Directoraat-Generaal Rijkswaterstaat, Meetkundige Dienst, The Netherlands, in Dutch.

Breunese, J., Mijnlieff, H. and Lutgert, J. (2005), The life cycle of the Netherlands' natural gas exploration: 40 years after Groningen, where are we now?, in: *Petroleum Geology: North-West Europe and Global Perspectives—Proceedings of the 6th Petroleum Geology Conference*, pp. 69–75, Geological Society, London.

Cassee, B. (2004), *Selection of Permanent Scatterer Candidates for Deformation Monitoring; Amplitude Calibration of ERS SLC SAR Images*, Master's thesis, Delft University of Technology.

CBG (1987), Studieresultaten betreffende ongelijkmatige zakkingen in verband met aardgaswinning in de provincie Groningen, Tech. rep., Commissie Bodemdaling door Aardgaswinning.

Chapman, R. E. (1983), *Petroleum Geology*, Elsevier, Amsterdam, 1st edn.

Chatfield, C. (1989), *The Analysis of Time Series*, Chapman & Hall, London, 4th edn.

Cheung, G., Prima, M. A., Maurenbrecher, P. M. and Schokking, F. (2000), Statistical analysis of benchmark stability prior to natural gas extraction in a holocene clay and peat area, province of Friesland, the Netherlands, in: *Proceedings of the Sixth International Symposium on Land Subsidence (SISOLS 2000), Ravenna, Italy, 24–29 September 2000.*

Colesanti, C., Ferretti, A., Novali, F., Prati, C. and Rocca, F. (2003), SAR monitoring of progressive and seasonal ground deformation using the permanent scatterers technique, *IEEE Transactions on Geoscience and Remote Sensing*, **41**(7):1685–1701.

Colesanti, C., Mouelic, S. L., Bennani, M., Raucoules, D., Carnec, C. and Ferretti, A. (2005), Detection of mining related ground instabilities using the permanent scatterers technique—a case study in the east of France, *International Journal of Remote Sensing*, **26**(1):201–207.

Craft, B. C. and Hawkins, M. (1991), *Applied Petroleum Reservoir Engineering*, Prentice-Hall Inc., Englewood Cliffs, 1st edn.

Cumming, I. and Wong, F. (2005), *Digital Processing of Synthetic Aperture Radar Data: Algorithms And Implementation*, Artech House Publishers, New York.

Dake, L. P. (2002), *Fundamentals of Reservoir Engineering*, Elsevier, Amsterdam, 19th edn.

de Bruijne, A., van Buren, J., Kösters, A. and van der Marel, H. (2005), *De geodetische referentiestelsels van Nederland*, Groene serie, Netherlands Geodetic Commission, Delft, in Dutch.

de Heus, H. M., Joosten, P., Martens, M. H. F. and Verhoef, H. M. E. (1994), Geodetische Deformatie Analyse: 1D-deformatieanalyse uit waterpasnetwerken, Tech. rep. 5, Delft University of Technology, LGR Series, Delft, in Dutch.

de Jager, J. and Geluk, M. C. (2007), *Geology of the Netherlands*, Royal Netherlands Academy of Arts and Sciences, Amsterdam.

de Loos, J. M. (1973), In situ compaction measurements in Groningen observation wells, *Verhandelingen van het Koninklijk Nederlands geologisch mijnbouwkundig Genootschap*, **28**:79–104.

de Mulder, E. F. J., Geluk, M. C., Ritsema, I., Westerhoff, W. E. and Wong, T. E. (2003), *De Ondergrond van Nederland*, Nederlands Instituut voor Toegepaste Geowetenschappen (TNO-NiTG), Utrecht.

De Zan, F. and Rocca, F. (2005), Coherent processing of long series of sar images, in: *International Geoscience and Remote Sensing Symposium, Seoul, Korea, 25–29 July 2005.*

DINO (2008), Geological databank of TNO-NiTG, Geological Survey of the Netherlands, www.dinoloket.nl.

Doornbos, E. and Scharroo, R. (2004), Improved ERS and Envisat precise orbit determination, in: *ENVISAT & ERS Symposium, Salzburg, Austria, 6–10 September, 2004.*

DORIS (2008), Doppler Orbitography and Radiopositioning Integrated by Satellite (DORIS), website of Centre National d'Etudes Spatiales (CNES), http://www.cnes.fr/web/1513-doris.php.

Duin, E. J. T., Doornenbal, J. C., Rijkers, R. H. B., Verbeek, J. W. and Wong, T. E. (2006), Subsurface structure of the Netherlands—results of recent on and offshore mapping, *Netherlands Journal of Geosciences—Geologie en Mijnbouw*, **85**(4):245–276.

Duquesnoy, A. J. H. M. (2002), Wettelijke voorschriften en normering bij de meting van bodembewegingen als gevolg van delfstoffenwinning, in: Barends, F. B. J., Kenselaar, F. and Schröder, F. H., eds., *Bodemdaling meten in Nederland. Hoe precies moet het? Hoe moet het precies?*, Netherlands Geodetic Commission, Delft, in Dutch.

Ferretti, A., Prati, C. and Rocca, F. (2000), Nonlinear subsidence rate estimation using permanent scatterers in differential SAR interferometry, *IEEE Transactions on Geoscience and Remote Sensing*, **38**(5):2202–2212.

Ferretti, A., Prati, C. and Rocca, F. (2001), Permanent scatterers in SAR interferometry, *IEEE Transactions on Geoscience and Remote Sensing*, **39**(1):8–20.

Ferretti, A., Perissin, D., Prati, C. and Rocca, F. (2005), On the physical nature of SAR Permanent Scatterers, in: *URSI Commission F Symposium on Microwave Remote Sensing of the Earth, Oceans, Ice and Atmosphere, Ispra, Italy, 20–21 April 2005.*

Ferretti, A., Savio, G., Barzaghi, R., Borghi, A., Musazzi, S., Novali, F., Prati, C. and Rocca, F. (2007), Submillimeter accuracy of InSAR time series: experimental validation, *IEEE Transactions on Geoscience and Remote Sensing*, **45**(5):1142–1153.

Fokker, P. A. (2002), Subsidence prediction and inversion of subsidence data, in: *SPE/ISRM Rock Mechanics Conference, Irving, Texas, 20–23 October 2002.*

Fokker, P. A. and Orlic, B. (2006), Semi-analytical modelling of subsidence, *Mathematical Geology*, **38**(5):565–589.

Fredrich, J. T., Arguello, J. G., Deitrick, G. L. and de Rouffignac, E. P. (2000), Geomechanical modeling of reservoir compaction, surface subsidence, and casing damage at the belridge diatomite field, *SPE Reservoir Evaluation and Engineering*, **3**(4):348–359.

Fruneau, B. (2003), Conventional and PS differential SAR interferometry for monitoring vertical deformation due to water pumping: the Haussmann-St-Lazare case example (Paris, France), in: *Third International Workshop on ERS SAR Interferometry, 'FRINGE03', Frascati, Italy, 1–5 Dec. 2003.*

Geertsma, J. (1973a), A basic theory of subsidence due to reservoir compaction: the homogeneous case, *Verhandelingen van het Koninklijk Nederlands geologisch mijnbouwkundig Genootschap*, **28**:43–62.

Geertsma, J. (1973b), Land subsidence above compacting oil and gas reservoirs, *Journal of Petroleum Technology*, pp. 734–744.

Geertsma, J. and van Opstal, G. (1973), A numerical technique for predicting subsidence above compacting reservoirs, based on the nucleus of strain concept, *Verhandelingen van het Koninklijk Nederlands geologisch mijnbouwkundig Genootschap*, **28**:63–78.

Grebenitcharsky, R. and Hanssen, R. F. (2005), A Matérn class covariance function for modeling atmospheric delays in SAR interferometry, in: *AGU Fall meeting, December 5–9, San Francisco, USA*.

Gruen, A. W. and Baltsavias, E. P. (1985), Adaptive least squares correlation with geometrical constraints, *SPIE Computer Vision for Robots*, **595**:72–82.

Hanssen, R. (2004), Stochastic modeling of time series radar interferometry, in: *International Geoscience and Remote Sensing Symposium, Anchorage, Alaska, 20–24 September 2004*.

Hanssen, R. and Usai, S. (1997), Interferometric phase analysis for monitoring slow deformation processes, in: *Third ERS Symposium—Space at the Service of our Environment, Florence, Italy, 17–21 March 1997*, ESA SP-414, pp. 487–491.

Hanssen, R. F. (2001), *Radar Interferometry: Data Interpretation and Error Analysis*, Kluwer Academic, Dordrecht.

Hanssen, R. F., Weckwerth, T. M., Zebker, H. A. and Klees, R. (1999), High-resolution water vapor mapping from interferometric radar measurements, *Science*, **283**:1295–1297.

Hejmanowski, R. and Sroka, A. (2000), Time-space ground subsidence prediction determined by volume extraction from the rock mass, in: *Proceedings of the Sixth International Symposium on Land Subsidence (SISOLS 2000), Ravenna, Italy, 24–29 September 2000*, pp. 367–375.

Hettema, M., Papamichos, E. and Schutjens, P. (2002), Subsidence delay: field observations and analysis, *Oil & Gas Science and Technology*, **57**(5):443–458.

Hoefnagels, A. A. J. V. (1995), Analyse van bewegingen van ondiep gefundeerde peilmerken boven het Groninger gasveld in de dertig jaar voorafgaand aan de gaswinning, Tech. rep., Memoir of the Centre of Engineering Geology in the Netherlands, no. 130, in Dutch.

Hoekman, D. H. and Quinones, M. J. (1998), Forest type classification by airborne SAR in the Columbian Amazon, in: *Second Int. Workshop on "Retrieval of Bio- and Geophysical Parameters from SAR Data for Land Applications", Noordwijk, The Netherlands, 21–23 Oct. 1998*.

Hooper, A., Zebker, H., Segall, P. and Kampes, B. (2004), A new method for measuring deformation on volcanoes and other non-urban areas using InSAR persistent scatterers, *Geophysical Research Letters*, **31**(23):L23611.1–L23611.5.

Houtenbos, A. P. E. M. (2004), *Subsidence Residual Modeling, SURE user manual*, A.P.E.M. Houtenbos Geodetic Consultancy.

Humme, A. (2007), *Point Density Optimization for SAR Interferometry; a Study Tested on Salt Mine Areas*, Master's thesis, Delft University of Technology.

Inglada, J., Souyris, J.-C. and Henry, C. (2004), ASAR multi-polarization images phase difference: assessment in the framework of persistent scatterers interferometry, in: *ENVISAT & ERS Symposium, Salzburg, Austria, 6–10 September 2004*.

Kampes, B. and Usai, S. (1999), Doris: the Delft object-oriented radar interferometric software, in: *2nd International Symposium on Operationalization of Remote Sensing, Enschede, The Netherlands, 16–20 August 1999*.

Kampes, B. M. (2005), *Displacement Parameter Estimation using Permanent Scatterer Interferometry*, Ph.D. thesis, Delft University of Technology.

Kenselaar, F. and Quadvlieg, R. (2001), Trend-signal modelling of land subsidence, in: *10th FIG, International Symposium on Deformation Measurements, Orange, California, USA, 19–22 March 2001*.

Ketelaar, G., Marinkovic, P. and Hanssen, R. (2004a), Validation of point scatterer phase statistics in multi-pass InSAR, in: *ENVISAT & ERS Symposium, Salzburg, Austria, 6–10 September, 2004*, 10 pp.

Ketelaar, G., van Leijen, F., Marinkovic, P. and Hanssen, R. (2005), Initial point selection and validation in PS-InSAR using integrated amplitude calibration, in: *International Geoscience and Remote Sensing Symposium, Seoul, Korea, 25–29 July 2005*, pp. 5490–5493.

Ketelaar, G., van Leijen, F., Marinkovic, P. and Hanssen, R. (2006), On the use of point target characteristics in the estimation of low subsidence rates due to gas extraction in Groningen, the Netherlands, in: *Fourth International Workshop on ERS/Envisat SAR Interferometry, 'FRINGE05', Frascati, Italy, 28 Nov.–2 Dec. 2005*, 6 pp.

Ketelaar, G., van Leijen, F., Marinkovic, P. and Hanssen, R. (2007a), Multi-track PS-InSAR datum connection, in: *International Geoscience and Remote Sensing Symposium, Barcelona, Spain, 23–27 July 2007*, 4 pp.

Ketelaar, G., van Leijen, F., Marinkovic, P. and Hanssen, R. (2007b), Multi-track PS-InSAR: datum connection and reliability assessment, in: *ESA ENVISAT Symposium, Montreux, Switzerland, 23–27 April 2007*, 6 pp.

Ketelaar, V. B. H. and Hanssen, R. F. (2003), Separation of different deformation regimes using INSAR data, in: *Third International Workshop on ERS SAR Interferometry, 'FRINGE03', Frascati, Italy, 1–5 Dec. 2003*, 6 pp.

Ketelaar, V. B. H., Hanssen, R. F., Houtenbos, A. P. E. M. and Lindenbergh, R. C. (2004b), Idealization precision of point scatterers for deformation modeling, in: *4th International Symposium on Retrieval of Bio- and Geophysical Parameters from SAR Data for Land Applications, Innsbruck, Austria, 16–19 Nov. 2004*, 8 pp.

Ketelaar, V. B. H., van Leijen, F. J., Marinkovic, P. S. and Hanssen, R. F. (2008a), Monitoring surface deformation induced by hydrocarbon production in Groningen, the Netherlands, submitted to *Journal of Geophysical Research*.

Ketelaar, V. B. H., van Leijen, F. J., Marinkovic, P. S. and Hanssen, R. F. (2008b), Multi-track PS-InSAR for deformation monitoring, submitted to *Remote Sensing of Environment*.

KODAC (2008), website of KODAC: Data Center of the Royal Netherlands Meteorological Institute (KNMI), http://www.knmi.nl/klimatologie/daggegevens.

Kwinta, A., Hejmanowski, R. and Sroka, A. (1996), A time function analysis used for prediction of rock mass subsidence, in: *Proceedings of the International Symposium on Mining Science and Technology, Xuzhou, Jiangsu, China, 16–18 October 1996*.

Landes, K. K. (1959), *Petroleum Geology*, Wiley, New York, 2nd edn.

Langbein, J. and Johnson, H. (1997), Correlated errors in geodetic time series: implications for time-dependent deformation, *Journal of Geophysical Research*, **102**(B1):591–603.

Laur, H., Bally, P., Meadows, P., Sanchez, J., Schaettler, B., Lopinto, E. and Esteban, D. (2002), Derivation of the backscattering coefficient σ^0 in ESA ERS SAR PRI products, Tech. Rep. ES-TN-RS-PM-HL09, ESA, Issue 2, Rev. 5d.

Lilliefors, H. W. (1967), On the Kolmogorov-Smirnov test for normality with mean and variance unknown, *Journal of the American Statistical Association*, **62**:399–402.

Liu, S., Kleijer, F. and Hanssen, R. F. (2008), Turbulence in the Earth's troposphere revealed by Synthetic Aperture Radar Interferometry, 9e Nederlands Aardwetenschappelijk Congres 18–19 March, 2008, Veldhoven.

Lutgert, J., Mijnlieff, H. and Breunese, J. (2005), Predicting gas production from future gas discoveries in the Netherlands: quantity, location, timing, quality, in: *Petroleum Geology: North-West Europe and Global Perspectives—Proceedings of the 6th Petroleum Geology Conference*, pp. 77–84, Geological Society, London.

Marinkovic, P. and Hanssen, R. (2007), Dynamic persistent scatterers interferometry, in: *International Geoscience and Remote Sensing Symposium, Barcelona, Spain, 23–27 July 2007*, 4 pp.

Marinkovic, P., Ketelaar, G. and Hanssen, R. (2004), A controled Envisat/ERS Permanent Scatterer experiment, implications of corner reflector monitoring, in: *CEOS SAR Workshop, Ulm Germany, 27–28 May 2004*.

Marinkovic, P., Ketelaar, G., van Leijen, F. and Hanssen, R. (2008), InSAR Quality control: analysis of five years of corner reflector time series, in: *Fifth International Workshop on ERS/Envisat SAR Interferometry, 'FRINGE07', Frascati, Italy, 26 Nov.–30 Nov. 2007*, 8 pp.

Marinkovic, P. S., Ketelaar, V. B. H. and Hanssen, R. F. (2006), Utilization of high-Doppler ERS acquisitions in interferometric time series, in: *European Conference on Synthetic Aperture Radar, Dresden, Germany, 16–18 May 2006*, 6 pp.

Massonnet, D., Rossi, M., Carmona, C., Adagna, F., Peltzer, G., Feigl, K. and Rabaute, T. (1993), The displacement field of the Landers earthquake mapped by radar interferometry, *Nature*, **364**(8):138–142.

Meisina, C., Zucca, F., Dossati, D., Ceriani, M. and Allievi, J. (2006), Ground deformation monitoring by using the Permanent Scatterers Technique, *Engineering Geology*, **88**:240–259.

Mijnbouwwet (2008), website of the Dutch governmental laws, http://wetten.overheid.nl.

Mogi, K. (1958), Relations between eruptions of various volcanoes and the deformations of the ground surfaces around them, *Bulletin of the Earthquake Research Institute, University of Tokyo*, **36**:99–134.

NAM (1991), Stabiliteitsanalyse Historie Peilmerken Groningen, Tech. rep., Nederlandse Aardolie Maatschappij B.V.

NAM (2003a), Opslagplan Norg, Nederlandse Aardolie Maatschappij B.V., http://www.nlog.nl.

NAM (2003b), Winningsplan Anjum, Nederlandse Aardolie Maatschappij B.V., http://www.nlog.nl.

NAM (2003c), Winningsplan Groningen, Nederlandse Aardolie Maatschappij B.V., http://www.nlog.nl.

NAM (2005), Bodemdaling door Aardgaswinning, Tech. rep., Nederlandse Aardolie Maatschappij B.V.

NAM (2006), Gaswinning onder de Waddenzee, Tech. rep., Nederlandse Aardolie Maatschappij B.V.

NAM (2008), Groningen Long-Term, Tech. rep., Nederlandse Aardolie Maatschappij B.V. and Stork GLT.

NLOG (2008), NL Oil and Gas Portal, website that provides information about oil and gas exploration and production in the Netherlands, managed by the Geological Survey of the Netherlands, www.nlog.nl.

Odijk, D. and Kenselaar, F. (2003), *Subsidence Modelling: user's manual of the SuMo software (version 4)*, Delft University of Technology.

Odijk, D., Kenselaar, F. and Hanssen, R. (2003), Integration of leveling and InSAR data for land subsidence monitoring, in: 11th *FIG International Symposium on Deformation Measurements, Santorini, Greece, 23–28 May 2003*, 8 pp.

Okada, Y. (1992), Internal deformation due to shear and tensile faults in a half-space, *Bulletin of the Seismological Society of America*, **82**(2):1018–1040.

Oppenheim, A. V., Willsky, A. S. and Young, I. T. (1983), *Signals and Systems*, Prentice-Hall International, London.

Perissin, D. (2006), *SAR Super-resolution and Characterization of Urban Targets*, Ph.D. thesis, Politecnico di Milano, Italy.

RDNAP (2008), website of the Dutch Geometric Infrastructure, maintained by the Kadaster and Rijkswaterstaat, http://www.rdnap.nl.

Rocca, F. (2007), Modeling interferogram stacks, *IEEE Transactions on Geoscience and Remote Sensing*, **45**(10):3289–3299.

Rodriguez, E., Morris, C. S., Belz, J. E., Chapin, E. C., Martin, J. M., Daffer, W. and Hensley, S. (2005), An Assessment of the SRTM Topographic Products, Tech. rep., Jet Propulsion Laboratory.

Rondeel, H. E., Batjes, D. A. J. and Nieuwenhuijs, W. H. (1996), *Geology of Gas and Oil under the Netherlands*, Kluwer Academic, Dordrecht.

Scharroo, R. and Visser, P. (1998), Precise orbit determination and gravity field improvement for the ERS satellites, *Journal of Geophysical Research*, **103**(C4):8113–8127.

Schoustra, S. S. (2004), Bodemdaling Groningen, Analyse van de waterpassingen 1964–2003, Tech. rep., Nederlandse Aardolie Maatschappij B.V.

Schoustra, S. S. (2006), Stabiele peilmerken Groningen, Onderzoek naar de stabiliteit van peilmerken in en rondom het Groningen gasveld, Tech. rep., Nederlandse Aardolie Maatschappij B.V.

Schroot, B. M., Bosch, J. H. A., Buitenkamp, H. S., Ebbing, J. H. J., de Lange, G., Lehnen, C., van der Linden, W. and Roos, W. (2003), Oorzaak schade aan gebouwen nabij Grou, Tech. rep., Nederlands Instituut voor Toegepaste Geowetenschappen (TNO-NiTG).

SCR (1993), in: *CEOS SAR Calibration Workshop, ESTEC, Noordwijk, The Netherlands, 20–24 Sept. 1993.*

SodM (2008), website of the Dutch State Supervision of Mines, http://sodm.nl.

Sroka, A. and Hejmanowski, R. (2006), Subsidence prediction caused by the oil and gas development, in: *3rd IAG Symposium on Geodesy for Geotechnical and Structural Engineering and 12th FIG Symposium on Deformation Measurements, Baden, Austria, 22–24 May 2006.*

SRTM (2008), Shuttle Radar Topography Mission website, maintained by Jet Propulsion Laboratory (JPL), http://www2.jpl.nasa.gov/srtm.

Teeuw, D. (1973), Laboratory measurement of compaction properties of Groningen reservoir rock, *Verhandelingen van het Koninklijk Nederlands geologisch mijnbouwkundig Genootschap*, **28**:19–32.

Teunissen, P. J. G. (1988), *Towards a Least-Squares Framework for Adjusting and Testing of both Functional and Stochastic Models*, vol. 26 of *Mathematical Geodesy and Positioning series*, Delft University of Technology.

Teunissen, P. J. G. (1995), The least-squares ambiguity decorrelation adjustment: a method for fast GPS integer ambiguity estimation, *Journal of Geodesy*, **70**(1–2):65–82.

Teunissen, P. J. G. (2000a), *Adjustment Theory; An Introduction*, Delft University Press, Delft, 1st edn.

Teunissen, P. J. G. (2000b), *Testing Theory; An Introduction*, Delft University Press, Delft, 1st edn.

Teunissen, P. J. G. (2001a), *Dynamic Data Processing; Recursive Least-squares*, Delft University Press, Delft, 1st edn.

Teunissen, P. J. G. (2001b), Statistical GNSS carrier phase ambiguity resolution: a review, in: *2001 IEEE Workshop Statistical Signal Processing, 6–8 August 2001, Singapore*, pp. 4–12.

Teunissen, P. J. G. (2007), Best prediction in linear models with mixed integer/real unknowns: theory and application, *Journal of Geodesy*, **81**:759–780.

Teunissen, P. J. G. and Odijk, D. (1997), Ambiguity dilution of precision: concept and application, in: *Proc. ION-97, 16–19 September, Kansas City, USA*, pp. 891–899.

Teunissen, P. J. G., Salzmann, M. A. and de Heus, H. M. (1987), Over het aansluiten van puntenvelden: De aansluitingsvereffening, *NTG Geodesia*, **29**(6/7):229–235/270–273.

Teunissen, P. J. G., Simons, D. G. and Tiberius, C. C. J. M. (2005), *Probability and Observation Theory*, Delft Institute of Earth Observation and Space Systems (DEOS), Delft University of Technology, The Netherlands.

Tiberius, C. C. J. M. and Kenselaar, F. (2003), Variance component estimation and precise GPS positioning: case study, *Journal of Surveying Engineering*, **129**(1):11–18.

Usai, S. (2001), *A New Approach for Long Term Monitoring of Deformations by Differential SAR Interferometry*, Ph.D. thesis, Delft University of Technology.

van der Kooij, M. (1997), Land subsidence measurements at the Belridge oil fields from ERS InSAR data, in: *Third ERS Symposium on Space at the Service of our Environment, Florence, Italy, 14–21 March 1997.*

van Leijen, F. J. and Hanssen, R. F. (2007), Persistent scatterer interferometry using adaptive deformation models, in: *ESA ENVISAT Symposium, Montreux, Switzerland, 23–27 April 2007*, 6 pp.

van Leijen, F. J., Hanssen, R. F., Marinkovic, P. S. and Kampes, B. M. (2006a), Spatio-temporal phase unwrapping using integer least-squares, in: *Fourth International Workshop on ERS/Envisat SAR Interferometry, 'FRINGE05', Frascati, Italy, 28 Nov–2 Dec 2005*, 6 pp.

van Leijen, F. J., Perski, Z. and Hanssen, R. F. (2006b), Error propagation and data quality assessment for ASAR persistent scatterer interferometry, in: *European Conference on Synthetic Aperture Radar, Dresden, Germany, 16–18 May 2006*, 6 pp.

van Meirvenne, M. and Goovaerts, P. (2002), Accounting for spatial dependence in the processing of multi-temporal SAR images using factorial kriging, *International Journal of Remote Sensing*, **23**(2):371–387.

van Zyl, J. J. (1989), Unsupervised classification of scattering behavior using radar polarimetry data, *IEEE Transactions on Geoscience and Remote Sensing*, **27**(1):36–45.

Verhoef, H. M. E., Joosten, P. and de Heus, H. M. (1996), Subsidence analysis in the Netherlands Groningen gasfield and the detection of (locally) unfitting points, in: *IAG Regional Symposium on Deformation and Crustal Movement Investigations using Geodetic Techniques, Szekesfehervar, Hungary, 31 August–5 September 1996*, pp. 159–166.

Verruijt, A. and van Baars, S. (2005), *Grondmechanica*, VSSD, Delft, 7th edn.

Wackernagel, H. (1998), *Multivariate Geostatistics*, Springer, Berlin, 2nd edn.

Welsch, W. M. and Heunecke, O. (2001), Models and terminology for the analysis of geodetic monitoring observations, in: *The* 10th *FIG International Symposium on Deformation Measurements, Orange, California, USA, 19–22 March 2001*, pp. 390–412.

Werner, C., Wegmuller, U., Strozzi, T. and Wiesmann, A. (2003), Interferometric Point Target analysis for deformation mapping, in: *International Geoscience and Remote Sensing Symposium, Toulouse, France, 21–25 July 2003*, 3 pp.

Wright, T. J., Parsons, B. E. and Lu, Z. (2004), Towards mapping surface deformation in three dimensions using InSAR, *Geophysical Research Letters*, **31**:5.

About the Author

Gini Ketelaar (1975, The Netherlands) studied geodesy at Delft University of Technology in the period 1993–1998. Her M.Sc. thesis focused on mathematical geodesy, more specifically on the testing procedure in the parameter estimation using geodetic measurements. For this thesis, she received the prize for best geodesy graduate 1998–1999. After obtaining her M.Sc. degree, Gini worked for Ingenieursbureau Geodelta B.V. as a project engineer (1998–2000), performing consultancy work and performing research in aerial and close range photogrammetry. In 2000, she started to work for Thales Geosolutions B.V., as an offshore surveyor in cable lay, multibeam echosounder, and geophysical surveying projects, worldwide. In 2003, she returned to Delft University of Technology for her PhD study, investigating the satellite radar interferometry (InSAR) technique for monitoring subsidence due to hydrocarbon production in the Netherlands. This research was performed in collaboration with Nederlandse Aardolie Maatschappij B.V. (NAM), where she is currently employed as geodesist in subsidence monitoring. Gini received her PhD degree at Delft University of Technology in September 2008. This book covers her findings on the application of InSAR for subsidence monitoring.

Index

GPSR Compliance

The European Union's (EU) General Product Safety Regulation (GPSR) is a set of rules that requires consumer products to be safe and our obligations to ensure this.

If you have any concerns about our products, you can contact us on ProductSafety@springernature.com

In case Publisher is established outside the EU, the EU authorized representative is:

Springer Nature Customer Service Center GmbH
Europaplatz 3
69115 Heidelberg, Germany

Batch number: 10755292

Printed by Printforce, the Netherlands